Spectacles and the Victorians

Manchester University Press

SOCIAL HISTORIES OF MEDICINE

Series editors: David Cantor, Anne Hanley and Elaine Leong

Social Histories of Medicine is concerned with all aspects of health, illness and medicine, from prehistory to the present, in every part of the world. The series covers the circumstances that promote health or illness, the ways in which people experience and explain such conditions, and what, practically, they do about them. Practitioners of all approaches to health and healing come within its scope, as do their ideas, beliefs, and practices, and the social, economic and cultural contexts in which they operate. Methodologically, the series welcomes relevant studies in social, economic, cultural, and intellectual history, as well as approaches derived from other disciplines in the arts, sciences, social sciences and humanities. The series is a collaboration between Manchester University Press and the Society for the Social History of Medicine.

To buy or to find out more about the books currently available in this series, please go to: https://manchesteruniversitypress.co.uk/series/social-histories-of-medicine/

Spectacles and the Victorians

Measuring, defining and shaping visual capacity

Gemma Almond-Brown

MANCHESTER UNIVERSITY PRESS

Copyright © Gemma Almond-Brown 2023

The right of Gemma Almond-Brown to be identified as the author of this work has been asserted by them in accordance with the Copyright, Designs and Patents Act 1988.

Published by Manchester University Press
Oxford Road, Manchester M13 9PL

www.manchesteruniversitypress.co.uk

British Library Cataloguing-in-Publication Data
A catalogue record for this book is available from the British Library

ISBN 978 1 5261 6135 2 hardback
ISBN 978 1 5261 9485 5 paperback

First published 2023
Paperback published 2026

The publisher has no responsibility for the persistence or accuracy of URLs for any external or third-party internet websites referred to in this book, and does not guarantee that any content on such websites is, or will remain, accurate or appropriate.

EU authorised representative for GPSR:
Easy Access System Europe – Mustamäe tee 50, 10621 Tallinn, Estonia
gpsr.requests@easproject.com

Typeset
by Cheshire Typesetting Ltd, Cuddington, Cheshire

For my grandma,
Joan Almond
1931–2022

Contents

List of figures	*page* viii
List of table	ix
Acknowledgements	x
Introducing Victorian spectacle wear	1
1 Early Victorian understandings of vision and spectacles, 1830–1850	39
2 The 'normal eye' as seen through technology: a quest for medical control, 1850–1904	80
3 Challenging (ab)normalcy: expansion in manufacture, design and access, 1851–1904	123
4 The limits of professionalism: medical practitioners, opticians and popular responses to sight loss, 1880–1904	163
5 Fashioning the eye and seeing, 1830–1904	207
Conclusion	246
Bibliography	257
Index	275

Figures

1.1	Trade card of A. Mackenzie.	49
1.2	Trade card of S. Phillips.	54
3.1	Transverse folding spectacles.	130
3.2	Extending spectacles.	131
3.3	Turn pin spectacles.	132
3.4	Straight spectacles.	133
3.5	Coil-spring spectacles.	134
3.6	Steel wire spectacles with additional material for comfort.	135
3.7	*Pince-nez*.	137
4.1	Trial lens case.	168
5.1	Rimless eyeglasses.	229
5.2	Decorative eyeglass frame.	230

Table

3.1 Average cost of different types of repairs from
John Potter Dowell and Robert Sadd's account books.   150

Acknowledgements

This book is the culmination of seven years of research which has greatly benefited from my experience at a variety of institutions and the contributions of many people.

First, I would like to thank David Turner. I walked into your office as a second-year undergraduate student who had just discovered the field of disability history. I am indebted to your mentoring, guidance and friendship over the last ten years, and I can safely say that this book would have never been written had we not met.

A large proportion of this research was undertaken between 2016 and 2018 in London at the Science Museum, as part of an AHRC award. I was incredibly fortunate to have had unlimited access to the Science Museum's object stores, but it was not without its challenges. A special thanks must go to Tim Boon for his patience and time in helping me to learn how to use, and communicate, object research in historical writing, and to Stewart Emmens for getting me to think like a curator for the first time. I am also grateful to everyone in the research collections team, the Dana Research Centre and Library staff and those at the off-site store in Wroughton for their expertise in helping me to locate relevant materials. I was fortunate to be a part of the research community there, with especial mention to Kevin, Ayesha, Harriet, Rachel, Laura, Stuart and Jacob. From training days, away days and coffee dates at the British Library and Wellcome Library I am just as grateful to the CDP scheme for the cross-organisational friendships of Hannah, Becky, Hannah and Anna.

I am indebted to Neil Handley, curator of the museum at the College of Optometrists. Our discussions, and your expertise, have been influential in framing the material in this book and I am grateful for unfettered access to the archives and objects at the museum, including the many, many volumes of *The Optician* that I got through. I am also grateful to the helpful input of archivists at the Boots Archives, Cambridgeshire Archives, Carlisle Archive Centre, Sheffield City Archives (a special thanks for posting my laptop's power lead that I left behind!), Trinity College Library at Cambridge University, University of Nottingham Archives Manuscripts & Special Collections, the Thackray Museum, Wellcome Library (especially Ross Macfarlane for his early encouragement) and the West Sussex Record Office.

Some very early drafts of this research were completed between 2018 and 2019 at the research office in Swansea University, which became a place of solace during some lengthy blocks of writing as well as a much-needed chance for social interaction. There are too many to name, but I'll at least try: Jay, Allyson, Rachel, Hillary, Brandi, Rio, Ben, Meilyr, Shareena, thank you for getting me through! I would also like to thank Graeme Gooday (Leeds University), Adam Mosely and Louise Miskell (Swansea University) for their insightful comments on and improvements to an early version of this work. Graeme, I am indebted to your comments on the narrative of medicalisation and normalcy that you identified across my chapters and helped me to bring to light.

I have presented material in this book at several conferences held by a variety of organisations, including the British Society for the History of Science, the Social History of Medicine and the Institute for Historical Research. The most influential for the framing of this book was the panel on normalcy organised by Coreen McGuire at the European Association for the History of Medicine in 2019. Coreen, thank you for inviting me to participate in numerous panels, putting me up in Leeds, organising various lunch and dinner dates (often with the additional, and welcome, company of Richard) and reading drafts of this book. I am especially grateful for your encouragement and friendship as well as your scholarship, which has greatly influenced, and inspired, my thinking.

The completion of this project has been made possible by a Royal Historical Society Early Career Researcher's Grant. In finishing this book, I am especially grateful to the anonymous reviewers.

You have made this research so much more robust and encouraged me to think more broadly about its significance. Special thanks must also go to Julie Anderson and Ryan Sweet for their feedback on chapters of this book and to Deborah Youngs, who has been a source of constant reassurance since 2011. I must also thank Sarah Crook for her friendship over the last three years in Swansea; I am in awe of your ability to balance your career and motherhood. Thank you for giving me the confidence to step back into academia after maternity leave and for your endless cheerleading in getting me to the end of this book.

This work would not have been completed without the love and understanding of my family. To my mum and dad, you never placed the limitations of medical diagnosis onto me. Thank you for instilling in me a firm belief that I could achieve anything I ever wanted to. Your parenting and guidance inspired my interest in the way the functional body is measured, and you taught me to highlight the ways in which narratives of medicalisation, normalisation and fixed categories of diagnosis fail to capture the reality, and should not obstruct the possibilities, of individual lived experience. To my sister Kate, thank you for laughing at the thought of my spending so many years of my life studying spectacles, but for supporting me through it anyway. The fact that you and Mark no longer need to feign interest when listening to my research is, for me, a success. To my grandparents – Kathleen Vernon, Robert Vernon and Peter Jones – I say I cannot wait to put this book in your hands. To my grandma Joan Almond, who sadly passed away as I finished this book, I know you were waiting for that day too and never lost the opportunity to tell everyone you met about it; I pushed through to finish this in your memory. Alison and Emma, this research would not have been possible if I had not had a loving home to stay in while I was researching in London; I will always have fond memories of that year and the Central Line. To the friends who have said they want a copy of the book to sit on their shelf (but not necessarily read) – Bryony, El, Libby, Al – thank you for always having my back. My family expanded during the writing of this book and as too did the support. Kim, Rich, Nathan, Georgie, Esther, Jesse and Sully, I know you don't choose your in-laws and nephews but I would choose you every time.

Finally, to Samuel and Joshua. Samuel, thank you for your belief in me, for knowing when I needed to stop (this would just have been constantly redrafted if it wasn't for you), and for gladly walking this journey with me. And to my son, Joshua, your birth in May 2020 was unplanned and unexpected, but it so greatly enhanced this work. Thank you for offering me the much-needed distance required to return to this with fresh eyes. You have given me a whole new perspective on life, your enthusiasm and your keenness to learn inspires me every day.

Introducing Victorian spectacle wear

A few years ago, I drove to my local optician's for a routine checkup and was surprised to be told that my vision had deteriorated to the point that the optometrist could not legally allow me to drive home without the use of spectacles. In what ways did my faded eyesight matter? How had I not noticed it nor found it personally problematic? As I tried to invoke memories of comparatively normal vision I started to wonder, compared to what norm? For the optometrist, my eyesight was a problem that was not measured in terms of my personal insights but through numerical standards carrying such weight that they could be used to place legal restraints on my activity. To understand this, we need to understand the history of spectacle usage in the Victorian period. It is only through analysis of Victorian developments in ophthalmology that we can understand where these normative vision standards have come from, how they are measured and how failures to meet them have come to be corrected through spectacles.

The Victorians did not invent spectacles. Spectacles had long been produced in artisanal contexts and were neither scarce nor expensive in Europe from as early as the fourteenth century.[1] By the seventeenth century, they were readily available across classes in Britain.[2] In the eighteenth century their access and utility was further enhanced by the broadening of materials used in spectacle production, the incorporation of spectacle side-arms and the emergence of desire for certain frames, all of which removed some of the prejudices associated with their wear.[3] The Victorians accelerated these trends. They established the parameters of appropriate vision, secured frames to the face unaided for the first time, mass-produced uniform styles that could be bought for a penny and fostered a growth in visible, public, usage. As the first full-length study of spectacles and vision testing, this book offers an innovative approach to studying technologies of the body in the past. As an everyday object governed by medical standards and principles of diagnosis to enhance a person's sensory capability, the collective study of spectacles and vision testing allows the histories of material culture, disability and medicine to be combined in new and fruitful ways. Such an approach not only expands our knowledge of spectacles and vision testing, but more broadly expands our knowledge of how medical ideas shaped understandings of bodily capacity, how an everyday or mundane object can inform historical work and the scholarly importance of the histories of technology, medicine and disability being in conversation with one another.

Victorians recognised a growth in spectacle wear and rooted its magnitude in specific changes to daily living. On 17 February 1885 Dr Dyce Davidson, professor of *materia medica* at Aberdeen University, addressed members of the Philosophical Society in the Bath Hotel on the influence of education on the eyesight of children. Davidson had developed his paper to respond to the pervasive belief that the nation's vision was deteriorating because of schooling and other evolving areas of nineteenth-century life, including the built environment and the rise of print. Davidson's summary stated:

> It is not many years ago… – even within my own memory – that spectacles were supposed to be necessary only to those of mature years, and indeed were looked upon as the first warning of declining age. A few also were compelled to wear them, chiefly those who were very short-sighted, but their numbers were so small that they were marked as peculiar and attracted in our attention as a person afflicted with a deformity of body or limb. Any young person wearing glasses was at once the object of remark.[4]

Davidson's observations are significant because he did not simply point to an increase in spectacle use. Rather, he recorded a broadening demographic of spectacle users. In arguing that spectacles were now worn by the young he suggested that the meaning of spectacles had changed; they were no longer associated with age, and wearing them was no longer considered 'peculiar' or akin to a deformity of body or limb. It is not possible to assess the truth behind Davidson's observations. Davidson himself apologised for not being able to give an 'exact' ratio, and instead drew upon the experience of an optician who had stated that his spectacle sales had increased ninefold in the last five years. In the absence of statistics on spectacle users, contemporaries highlighted the palpable increase in spectacle use in a range of popular print to raise alarm about the deterioration of vision on a national scale. The role of medical practitioners, and more specifically ophthalmologists, in promoting, and engaging with, this discourse is striking.

This book attempts to explicate the relationship between the associated increase in spectacle use and the particularities of Victorian society. Why did Victorians, like Davidson, perceive the growth in spectacle use to be dramatic? Who wore spectacles? What processes allowed an increase in spectacle use to take place and, indeed, did use in fact increase? How did changes in spectacle wearing affect their social and cultural meanings and contemporaries' reactions to them? How did they transform from a peculiar assistive device to a mainstream commodity? What was the role of medical practitioners in this process? Medical adoption of technology was vital in recategorising contemporary understandings of vision and spectacles. At the beginning of the nineteenth century, how spectacles worked was ascribed to abstract, disembodied theories of refraction. Spectacles were not part of medical practice and if medically qualified personnel considered their patient's vision to be deteriorating or weak, they would refer them to an optician. The professional boundaries between the two were clear-cut. The optician was acknowledged as a specialist who understood the refractive properties of glass and how it could be applied to the eye. Their prospective customers were not particularly diverse; 'weak vision' described a broad category of spectacle users which primarily consisted of the elderly and a minority with severe short sight. The incorporation of diagnostic and vision testing equipment from the 1850s altered contemporaries' anatomical understanding of vision and optics, and their possibilities. In particular, the ophthalmoscope – and its subsequent adaption and application – allowed its many operators to gain a much fuller understanding of the eye which, in turn, expanded the number of conditions spectacles were thought capable of remedying.

To answer my earlier questions and investigate a period of rapid change I have two principal lines of enquiry: how was vision increasingly defined, measured and categorised through the adoption of new technology and why were medical practitioners unable to exert their control over spectacle use at a time of medical professional expansion? Crucially, ophthalmologists were only partially successful in their attempts to utilise technology to leverage their position and gain a monopoly over vision testing and spectacle dispensing. The number of spectacle users exceeded the capacity of medical practice and spectacles, as a device, came to embody a range of different meanings that prevented them from being viewed solely as a medical treatment device. The modification of spectacle use by ophthalmologists, designers and users was intimately intertwined with broader nineteenth-century and Victorian processes of change. While ophthalmologists asserted their position at a time of medical expansion, specialisation and professionalisation, the realities of spectacle use form a unique case study for exploring the limits, legitimacy and credibility of medical authority.

This book therefore gives equal weight to how and why medical practice changed and how this change was received and filtered into practice. By analysing spectacles as an assistive device and an accessory of display, I offer new insights into the process of medicalisation in the nineteenth century. Despite the emergence of the hospital as a site of spectacle dispensing, opticians and a variety of retailers fought to maintain a legitimate position on the high street, and succeeded. This has influenced the framing of the book, which begins in 1830 with retailers increasing alignment with the medical profession and finishes in 1904, when the Spectacle Makers Company included vision testing in its examinations. I connect this analysis of medicalisation, and its limits, to Victorian understandings of disability and the measurement of bodily capacity. Medical practitioners utilised technology to standardise and categorise vision against a new 'norm' which, in turn, shaped the definitions and experience of vision loss. On one hand, spectacles were able to 'restore' vision for some previously considered blind, and on the other hand, their advocates diagnosed a growing proportion of people unable to meet 'normal' vision requirements who had not previously thought they required treatment. Medical attempts to control and define the treatment of visual acuity, coupled with subsequent conflict and eventual collaboration with traders, adds to growing scholarship on the existence, and complexities, of Victorian medical capitalism.

As a social history of medicine, this book draws upon the social and cultural change that occurred in tandem with transforming medical practice. The rise of print and visual culture, education and the built environment served to reinforce the importance of vision and the practical value of spectacles as a device and a commodity. Spectacle usage diversified in conjunction with a broadening user demographic that acknowledged their restorative properties. Practically, the ability to fix spectacles behind the ears was a key design challenge that was overcome in this period, thus enhancing the utility of spectacles for an increasing range of daily activities, whether those be riding a horse or reading a paper. However, spectacles equally embodied a range of meanings not associated with their practical function and the concept of fashion and fashioning of the body is just as important in informing this analysis. Spectacles emerged on display in the broadening thoroughfares of towns and cities, at the theatre, the races, the workplace and social events. The number of spectacle users partly explains the limits of medicalisation, as well as the array of cultural meanings that developed around spectacles. Other concerns, for example, those surrounding worker safety and workers' bodies, simultaneously informed, bolstered and challenged medical definitions and diagnosis. Spectacles, as I will argue, are vital for developing our understanding of Victorian medical change and responses to broader Victorian cultural change.

Spectacles, vision and the Victorians

The use and adoption of spectacles in the Victorian period has received little scholarly attention. Historical works have predominantly focused on the early history of visual aids and unearthing the broader history of spectacles has been primarily undertaken by private collectors. Collectors' work has provided the crucial foundations for any study of spectacles, including information about the manufacture, design and use of spectacles and lenses. From these studies, it is possible to know what designs or materials were popular and when and who invented certain styles. However, the prioritisation of certain objects by collectors, and the lack of acknowledgement by historians of the benefits of collectors' work and approach in this area, has limited historians' interest in the topic. A number of articles on the history of visual aids in the *Ophthalmic Antiques Collectors Club Newsletter*, for example, informatively draw on material such as trade literature to provide the initial ingredients required to begin compiling a full history of Victorian spectacles.[5] Yet collectors and historians, often answering and exploring different questions, have often remained separate in their endeavours and little attention has been paid to how this existing body of work could be further developed. In particular, the tendency of collectors' accounts to focus on how surviving objects illustrate key changes in design is invaluable to the historian who is interested in placing these objects in their broader social, cultural and medical context, and in unearthing the experiences of their past users. Some isolated attempts to situate spectacles and eyewear in relation to broader patterns of historical change reveal the potential of such a study for the Victorian period. They demonstrate that spectacles can be used to inform our historical understanding of a range of different topics, including scientific-instrument making and manufacture, the experiences of childhood, fashion and style, marketing and branding, of gender, and the historical relationship between technology and the body.[6]

At first glance, it seems surprising that the historian's attention has not hitherto been directed to the nineteenth and early twentieth centuries in light of the value of spectacles as an analytical tool and the significance of this period for shaping spectacle use. The Victorians privileged vision and this provides a diverse, complex and expanding cultural context in which to frame spectacles. Nevertheless, the most detailed existing work is limited to a chapter on spectacles, cameras and vision as part of a more comprehensive study of 'Victorian things'.[7] A full-length study of spectacles has relied on recent developments and changes within the historical discipline itself. Historical inattention can be partly explained by a lack of crossover between collectors' and historians' work, but it is also a product of historians' relatively late acknowledgement of the value of material culture. A full-length study of spectacles requires the study of objects to fully explain changes in design and usage. In addition, the study of a more ambiguous, and miscellaneous, assistive device forms part of a newly expanding area of history – disability and prosthetics – one that embraces history from below and popular lived experience.

Vision has been celebrated for centuries. More recently, its elevation has been critiqued and phrases such as the 'crisis in ocular centrism' have been coined to emphasise the danger of monosensory studies.[8] Historians such as Graeme Gooday and Karen Sayer have challenged both the importance of vision and its centrality in the past by looking at the experiences of sensory loss and multiple sensory loss.[9] However, we lack studies that explain this sensory hierarchy and it is therefore crucial to analyse the historical grading of the senses through a monograph on monosensory loss, especially one that explains the cultural value of vision. This is especially pressing given that blindness has been given less attention than deafness in recent years.[10] Sensory experience is not transhistorical. The historical investigation of the senses is 'not a study of biological universals but of the cultural attitudes that constitute and accompany perception'.[11] To understand the use and adoption of spectacles in the Victorian period, we need to understand how vision was perceived by the Victorians. Vision was celebrated during that period: but why? Building upon the work of Jonathan Crary, Chris Otter has argued that sensory perception, and a 'modern' way of seeing, became increasingly rooted in physiology. Vision's role and status profited from this positioning. In Victorian culture vision was doubly celebrated because it could be separated into two specific experiences: sensation, defined as a physiological process, and perception, defined as a psychological one. Vision was therefore valued because of its roles in acquiring knowledge and judging 'proofs' (i.e. perception), and because the eye was perceived as perfect in design and function (i.e. enabling vision).[12] This 'hierarchy of the senses' proliferated in a popular culture that was fascinated with the processes of seeing and with visual and ocular spectacle.

Vision was further cemented at the top of a hierarchal order of the senses because it was increasingly considered necessary for functioning in the Victorian world. A range of earlier studies have noted the uniqueness of the nineteenth-century environment for galvanising and expanding the importance of vision in conducting everyday activities. Within urban space, vision was given primacy. Street signs, house numbers and street lighting demanded an unprecedented level of visual acuity.[13] The effects of a changing environment were two-fold: it elevated both the importance of vision and the problem of vision loss or partial sight. Important work in sensory history has tracked how the thresholds or categorisation of sense perception are specific to time and place.[14] But Martin M. Smith, in reflecting on the current field of sensory history, argues that more needs to be done to consider 'a broader variety of sensory constituencies'.[15] In this book I connect this demand with recent trends in disability history that seek to explore how sensory, physical and mental capacities are classified. As outlined by David M. Turner in his study of disabilities and emotion, 'hierarchies of the senses' produce 'a hierarchy of impairment'.[16] Individual experience could be varied and, in nineteenth-century comparisons between the experiences of deaf and blind people, deafness could be considered a 'deeper and more complex' 'problem'.[17] Nevertheless, popular print often drew upon comparative sensory loss to explain the greatly magnified suffering or restrictions that would follow from either losing or injuring sight. Why was this the case and did Victorian reflections on the 'problem' of blindness evolve? Previous historical work on Victorian blindness does not answer these questions and has tended to be more institutional in focus, placing emphasis on the evolution of charity, education or organisations for the blind.[18] The scope of this book is different. I do not intend to explore the traditional 'blind' subject. Instead, I conceive of Victorian partial sight as part of a much broader spectrum of visual acuity, one that directly interacts with, and is influenced by, broader social change, the introduction of technology and medical ideas. Concerns about the condition of Victorian vision accelerated alongside theories, proposed by multiple actors, that the nation's vision was deteriorating in, and because of, the very environment that promoted it.

Victorians were faced with a conundrum: the problem of vision was rooted in the urban space and cultural change that had urged its importance. In 1892 an article in a monthly periodical reflecting on 'some social changes in fifty years', argued that many 'inestimable benefits' must be set against 'the increase of spectacle wearers and other indications of a decidedly lower sight average'.[19] The products of civilisation, and the development of towns and city centres, were also referred to as part of this commentary on social change. A variety of protective visual aids developed in response to transforming pastimes and transport methods that required participants to guard their eyes from wind, dust or bright light. In 1890 an advertisement from London-based opticians Thomas Harris & Son highlighted that there were frames suited to a range of protective functions, as well as physical activities:

> THOS HARRIS and SON'S SPECTACLES for BICYCLISTS
> THOS HARRIS and SON, SPECTACLES for CYCLISTS
> THOS HARRIS and SON, SPECTACLES for LAWN TENNIS
> THOS HARRIS and SON, SPECTACLES for BILLIARDS
> THOS HARRIS and SON, SPECTACLES for SHOOTING
> THOS HARRIS and SON, SPECTACLES for CRICKET
> THOS HARRIS and SON, SPECTACLES, from 3s
> THOS HARRIS and SON, SPECTACLES for PROTECTION from SUN
> THOS HARRIS and SON, SPECTACLES for PROTECTION from DUST
> THOS HARRIS and SON, SPECTACLES for PROTECTION from WIND
> THOS HARRIS AND SON, OPTICIANS, 32 GRACECHURCH-STREET.[20]

Deskbound pursuits also necessitated the use of lenses to remedy the refractive capacity of a person's eyes. Alongside the impact of the enclosed streets of cities and towns, and the changing methods of transport that demanded protection from the elements, the Victorian public drew attention to the effects of a more sedentary lifestyle on the population's vision. Particular attention was paid to the form and methods of education and the proliferation of print, which had led to an increase in reading for leisure. The problems associated with sedentary pursuits and the urban environment could be viewed collectively. Parallels can be drawn between this and criticism of the primacy of vision in city and modern environments by twentieth-century European thinkers who argued that it created a social passivity – people looked, rather than communicating.[21] Moreover, Peter John Brownlee's work on antebellum America – a comparable study on the emergence of an American 'ocular age' – has identified the role of print and visual culture in developing an 'economy of the eyes'. Culture targeted the eyes, placed newfound demands on visual acuity and had economic or personal repercussions for those unable to meet the visual requirements for their work or leisure. Vision, and its subsequent restoration through spectacles, became a sought-after commodity.[22]

Neither study of Victorian spectacles nor study of Victorian vision have fully explored how or why the increase in spectacle usage occurred at a time of heightened anxiety about the fallibility, and importance, of vision for everyday functioning. Victorians attributed their declining vision to the increasingly artificial nature of the Victorian built environment and leisure pursuits. Several comparisons were made in popular literature, for example, between Western societies and animals or indigenous populations to conclude that vision loss was both promoted and suffered as a result of developing civilisation.[23] Scholars have accordingly noted that spectacles were celebrated as a means to remedy the negative implications of civilisation in Britain at this time. However, the growth in spectacle usage cannot be generalised or explained by focusing only on the heightened role of vision. While Brownlee's study does not extend to the implications of medical advances in the 1850s, he rightly acknowledges that contemporaries needed to understand the function of spectacles. This was a two-way process that involved both changes to medical theory and practice, and increased attempts to disseminate this knowledge among the general public.[24] A study of the Victorian period is well placed to explore the effects of medical intervention on spectacle usage in greater detail. Unlike previous studies, I also argue, first, that the growth in spectacle wearing relied on alterations in spectacle design and innovations in manufacture; and second, that spectacle usage, and its distribution, needs to be explored in the context of the device's functionality and its perception and assimilation within cultural practice, a process that cannot be necessarily explained by the heightened value of, and/or concerns about, vision. By taking these areas into consideration, explaining the increase in spectacle usage allows the value and problem of Victorian vision to be interrogated from a new perspective. Spectacles served to reinforce and problematise vision and its position. Spectacle design, manufacture and dispensing, medical and popular understanding of the utility of visual aids and the experience of visual aid wearers are all vital interactions in determining the incidence and assimilation of spectacle wearing.

Interrogating medicalisation

A focused study of spectacles is both enriched by, and contributes to, our understanding of the history of Victorian society and medicine. The involvement of ophthalmologists in shaping the diagnosis of refractive vision errors and the adoption of lenses as a treatment method suggests, in many ways, that vision and partial sight became medicalised. Such an argument is further bolstered by the timing of this medical intervention; the measurement and testing of vision occurred in conjunction with the growth in ophthalmology as a medical specialty. A number of historical studies have focused on the specialisation of medicine in the nineteenth century more broadly, and they position ophthalmology at the forefront of a range of other disciplines, including obstetrics, paediatrics and orthopaedics.[25] Yet, outside this institutional focus, little work has connected the earlier advances in ophthalmology to the later developments in optometry.[26] This portion of history has remained more firmly within the profession and isolated from broader discussions of medical specialisation and medical authority.[27] I argue that optometry was a contestable feature of ophthalmologists' practice; simultaneously bolstering its narrative of success and acting as a professional Achilles' heel. On one hand ophthalmology, and the interest in the eye that this discipline helped to generate, influenced advances in determining, diagnosing and attempting to remedy a range of refractive vision errors. For example, while a number of historians have concluded that the invention of the ophthalmoscope – which enabled an individual to look inside a person's eye – was the most important development in this field during this period, they have done so without exploring how medical practitioners used the instrument to define new medical terms and states of the eye.[28]

Ophthalmologists transformed how the refractive and accommodative condition of the eye was understood and altered the way that spectacles could be used in the 'treatment' process. The adoption of the ophthalmoscope and other technologies demonstrate links of significance that can be made between developments in optometry and general medicine. In both, for example, greater emphasis was being placed on closer examination and diagnosis being performed by practitioners, rather than relying on the patient's account.[29] The hospital as an emergency site for spectacle dispensing also followed broader trends in the development of a range of specialised medical institutions. On the other hand, a more focused study of optometry within the discipline of ophthalmology exposes broader problems relating to professionalism, regulation and quackery in medicine at large.[30]

The diagnosis of the refractive and accommodative state of the eye may seem a model example of the process of medicalisation described by Michel Foucault, because in this way medical practitioners created a medical problem that had not previously been there.[31] However, it does not fully explain the medicalisation of the Victorian eye nor more recent scholarly treatment of the concept. Medicalisation is about defining a mental/bodily condition or state as a medical pathology, regardless of whether it is seen as a problem by those who are not medical professionals. The new medical pathology – in this case refractive and accommodative vision errors – is then subject to the expertise of medical practitioners who determine medical solutions or treatment, regardless of whether a 'cure' is required or the evidence of its medical nature is tenuous or dubious. Ophthalmology exposed and encouraged discussion of the fallibility of the eye and, in drawing upon the discourse of vision's importance, emphasised the overall value and, by implication, the severity of the problem of partial sight. This, in turn, led to discussion of the built environment in relation to bodily health, physical and mental limits and, in particular, eye strain. Vision defects could be and were problems for those who had differing severities of partial sight before these differences were fully recognised as such by medical practitioners. However, the creation of a 'normal' standard of vision, and medical methods of diagnosis, were more fundamental to altering contemporary understandings of the eye and its capacity.

Peter Conrad notes the importance of definitions for understanding the concept of medicalisation, the process of medicalisation and the legitimacy of expanding medical jurisdiction.[32] Definitions and diagnosis were central to ophthalmological practice. The refractive state of the eye – for example 'myopia' or short-sightedness – became either a disease or a dysfunction to be diagnosed according to changing medical understandings of health and pathology, precisely as a consequence of being named and defined.[33] Thus, the newly defined state of a person's eye became a 'vision error' or 'defect' that required medical treatment or cure.

While at first glance Victorian definitions and measurement of the eye provide a potentially strong example of the process of medicalisation, it is also a model or illustrative case for exposing the limits of this process. It is important to explore advances in medical knowledge and attempts to medicalise spectacles at both the elite and popular levels to fully characterise this trend. Although it has long been recognised that developments in new understandings of the eye occurred, the extent to which they affected the average user has not previously been explored. It cannot be presumed that these advances were automatically or immediately adopted in practice. Mary Carpenter, for example, is sceptical of the extent to which elite Victorian medical knowledge filtered into practice, and the speed with which it did so, and has argued that more research is required to assess how ideas were distributed.[34] In discussing the retailing, design and use of spectacles, Chapters Three to Five will highlight the limits to medical authority in their attempts to gain a monopoly over the understanding and dispensing of spectacles. Challenges to ophthalmologists' expertise and their newly defined categories of vision expose the potential illegitimacy of some aspects of Victorian medical practice, including its contested jurisdictional boundaries and the rise of nascent professions.[35] Ophthalmology, although the first medical specialty, had a long history of contending with quackery and a range of itinerant 'oculists' and 'experts'. As they became increasingly more involved in optometry ophthalmologists faced a new kind of competition: the multifarious spectacle dispenser.

Ophthalmologists and retailers emerged as competitors in a vast and complex visual aid marketplace. This marketplace offers a new perspective on the broader Victorian medical marketplace and the role of prosthetics within it. Between 1880 and the early twentieth century the Victorian medical marketplace expanded to include a variety of patent medicines, pharmaceutical remedies, technologies and assistive devices intended to improve a person's health or alter their sensory or physical capacity.[36] Scholars have demonstrated that traditional approaches to the medical marketplace have not adequately foregrounded the role of the medical practitioner either as a producer or as a consumer of medical products within this commercial landscape.[37] As outlined by Takahiro Ueyama, the relationship between medical practitioners and the commercial market was often one of forced collaboration rather than conflict.[38] Indeed, more recent scholarly work has identified that the commodification of prostheses was part of the broader process of asserting medical control over people's bodies.[39] For spectacles and eyeglasses, this was not the case. By the late Victorian period, visual aids are peculiar and function in the medical marketplace differently to, for example, the steel truss, artificial limb or hearing aid.[40] As Chapters Three to Five will demonstrate the variety of visual aid wearers meant that certain designs were mass-marketed, and their use and sale were influenced by a variety of forces outside medical control: a mass market that practitioners were, grudgingly or not, forced to accept and engage with.

The extent and limitations of medical influence on distribution and access of technologies of the body is a key measure of medicalisation. Conrad and Robert A. Nye, for example, have demonstrated that medical authority and jurisdiction in this area was a requirement for the growth and development of the medicalisation of society.[41] As will be seen by comparing Chapter One with Chapters Two and Four, the growth of medical authority in the study of the eye influenced how spectacles were dispensed and, by implication, the extent to which they were perceived as a medical object. However, ophthalmologists were unable to fully subjugate retailers, or to prevent a multitude of popular opinions that did not perceive spectacles in medical terms and therefore subsequently limited the legitimacy of medical claims. Although ophthalmologists designed ophthalmoscopes and diagnostic equipment, they relied on opticians and instrument makers to produce these tools to treat the range of refractive vision errors that they diagnosed. The sale and use of spectacles therefore created a competing, but increasingly collaborative, space between medical practitioners defining and shaping new theories of vision, the high street retailers and wholesalers that possessed the ability to manufacture and supply spectacles and equipment, and the public who chose and wore spectacles on their own terms. Medicalisation was still influential, but in ways that challenged its very nature; it shaped new professional boundaries by leading to the education and certification of opticians and generating a more enhanced awareness of vision and its refractive capacity among the general population.

Assisting the body: technology and the creation of functional visual norms

Scholarly work on the ways in which medicine can have a far-reaching influence on the use of prosthetic technologies and on contemporary understandings of bodily capacity, bodily limitations and bodily norms is well established. Is it appropriate to categorise spectacles as a 'prosthetic' or 'assistive device' and include them in this body of work? Is the condition of not being able to see clearly, unless wearing spectacles, a disability or an impairment? The history of prosthetics has expanded in recent years to incorporate a multitude of assistive devices, ranging from amplifying technology and dentures to the walking stick. These devices, which expanded from the eighteenth century onwards, are difficult to categorise and their definitions evolve. Spectacles, for example, have been defined as an orthotic, assistive device and prosthetic, and multiple definitions have been used by the same author across time.[42] Katherine Ott, more recently, has defined 'eyeglasses' as a 'prosthetic'. Within this definition, she places 'prosthetics 'within the broad category of assistive devices that people use to support what they want to do' and, specifically, that enhance 'mobility and agility, sensory apprehension, communication and cognitive function'.[43] The World Health Organization in 2015, as part of the Global Cooperation on Assistive Technology, defined an 'Assistive Product' in similarly broad terms as equipment, devices, instruments or software 'especially designed and produced or generally available, whose primary purpose is to maintain or improve an individual's functioning and independence and thereby promote their wellbeing'.[44] Assistive devices are not always perceived as prosthetics in the strict sense of the word – they do not replace a missing body part – but, as Ott recognised, they deepen our understanding of the themes that have emerged in the more established study of prosthetic technology and the body.[45] I argue that spectacles, and the processes adopted to identify the type and strength of a person's optical lens, provide a unique case study capable of revealing the multifaceted ways in which technology shapes the conceptualisation of disability and impairment. To understand my argument, we need to see through the eyes of a Victorian. How Victorians viewed their world, and what they expected their visual range to be, are different from today and were influenced by the varying efficacy of the technologies used to mediate or enhance their sensory capacity. Prior to the adoption of spectacles, in the early and late Victorian period those with the refractive conditions hypermetropia (more commonly known as long sight), myopia (more commonly called short sight) and astigmatism often considered their visual state to be 'blind'. Indeed, both patients and medical practitioners perceived their condition, and its associated symptoms, as 'disabling'.

The history of blindness has not considered visual aids in the context of a broader, more diverse spectrum of impaired visual acuity. As scholars have noted, blindness is difficult to define. Although the state of blindness became increasingly indexed and quantified in the nineteenth and twentieth centuries, there remains no single, unified, medical definition.[46] I argue that, as a subjective state, the experience of blindness and partial sight changed in the Victorian period in response to more thorough measurements of visual acuity and improved knowledge of the eye and the use of optical lenses. Visual aids therefore serve to highlight how the experiences and shaping of disability depend on, and can be defined by, the availability or utility of assistive devices. Work on assistive hearing aids, for example, has similarly revealed how their adoption reconceptualised what it meant to be 'deaf' in a given period.[47] Such findings illustrate that disability is not a transhistorical state across time, or even within a person's own lifetime. Disability is deeply contextual and dynamic.[48] Indeed, current legislation does not confine its definition of disability to one that is life-long or permanent and instead considers disability to encompass 'a physical or mental condition which has a long-term and substantial effect on your daily life'.[49] For the Victorians, whether because of cost, complexity of condition or lack of understanding or access to an appropriate dispenser, partial sight had a long-term, substantial effect on their daily lives. The adoption of new or improved diagnostic and assistive technology altered, and significantly reduced, these effects. Visual aids alleviated the symptoms of sight loss and in turn created the modern category of blindness as a totalising state of sight loss incapable of amelioration through technological means.

How technology shapes our understanding of disability and impairment is a thread throughout this book, finishing with the ways in which it ultimately shaped the experiences of its users in Chapter Five. Technology was vital in the classification of Victorian vision. Scholars have long noted the role of statistics and numbers in the creation of nineteenth-century categories of normalcy. Lennard J. Davis's work on hearing and deafness has been particularly influential in highlighting how statistics reconceptualised bodies that did not conform to the normal as deviant, abnormal problems that required correction.[50] Technology was not absent from Davis's work; prosthetics were the solution that enabled their users to conform to newly ascribed standards. But he pays little attention to the role of technology in shaping the very definitions of normalcy and abnormalcy. Coreen McGuire's more recent work demonstrates the fundamental role of diagnostic technology in defining and measuring normalcy and thus shaping understandings of bodily capacity during the period between the First and Second World Wars.[51] In this book, I highlight the similar role of technology and Victorian medical practitioners' perceived importance of objectively measuring the body. Technology and statistics both standardised and normalised vision from the 1850s onwards. In 1864, the seminal work on vision by Dutch ophthalmologist Franciscus Cornelis Donders utilised the ophthalmoscope to stimulate the discussion of vision within this framework.[52] While the emmetropic eye was 'perfectly formed' and normal, the ametropic eye was imperfect and abnormal. As I will argue, if vision testing tools were objective, the way Victorians interpreted the findings gleaned from their use was not.

In considering how technology and statistics shaped contemporary understandings of the Victorian body, however, care must be taken to not ascribe too much weight to the fixed regulatory power of the 'norm'. As Peter Cryle and Elizabeth Stephen's recent study argues, the 'normal' has been treated uncritically as oppressive; an uncontested power capable of defining people as 'pathological' or 'deviant'.[53] They also argue that present-day understandings of the normal are a twentieth century phenomenon and that in the nineteenth century the term was a scientific concept, divorced from understandings of bodily health.[54] Such a statement is perhaps too general; a standard for vision was accepted in the nineteenth century, and the concept of 'normal' vision did influence how a person's ocular and more general systemic health was perceived. But, nevertheless, I too seek to limit the norm's epistemic scope. This book exposes the fractured foundations, and the contested thought processes behind, 'normative vision'. This is not to underestimate the power of language in defining a person's lived experience. The concept of 'normal' vision did have an effect and could be used to exclude people from work environments. However, it is just as important to explore how normative vision emerged amid conflicts of interest and exposed the individuality of people's sensory capacity.

If we take inability to see as disabling in the Victorian period, and the subsequent, relative success of spectacles and technological medical intervention, am I suggesting that technology can 'cure' a disability or reduce an impairment? Assistive technologies certainly do have the capacity to standardise and shape the parameters and meaning of the functional body.[55] Spectacles and their utility defined contemporary parameters of functional visual capacity and standardised vision by determining the degree to which it should be 'corrected'. Here, spectacles are often considered unique. Studies of other assistive devices, for example, compare the relative success of spectacles in 'improving' vision to the restoration of function to other parts of the body.[56] Unique as spectacles might be, we do need to tread carefully on two accounts. First, this is not a whiggish narrative of progress. Victorian spectacles were not wholly 'restorative', and their success was not predetermined. Several users struggled to find lenses able to remove a host of associated symptoms that left them incapacitated, including blurred vision, fatigue, nausea and headache. Second, 'curing' disability is a goal considered to be part of an out-dated medical model of disability, a model that places emphasis on a person's bodily difference as a 'problem' that needs to be 'fixed'. However, while we should rightly critique this method of thinking, we should not disregard the medical model in its entirety, or its influence. Recent studies have demonstrated that the lack of previous conversation between the histories of medicine, technology and disability has limited our ability to interrogate the processes of medicalisation in defining and shaping people's experience.[57] Such analysis can, in fact, bolster alternative models of disability by exposing the limits of the medical model. In the case of Victorian vision, contemporaries did diagnose and define any difference in the refractive condition of the eye as an 'error', 'abnormality' or 'disease' but these terms were contested and not passively adopted by the population at large. Such an approach also can identify that medical 'problems' might have been 'problems' before they were defined in medical terms. Intervention could, in some cases, have newfound benefits; enhanced vision enabled an unprecedented number and range of spectacle users to maintain their vision for work and a variety of leisure pursuits. I therefore opt, where possible, to use the term 'enhancing' – 'correcting' and 'treating' appear only when quoting or discussing the ideas of Victorian contemporaries – in order to distance my work from the medical model and the idea of 'cure' while also acknowledging the subjective state of blindness and the need to interrogate the medical and optical intervention taking place.

Confronting the processes of medicalisation and medical terminology in shaping meanings of disability circumscribes, rather than widens, its scope. Vision was not completely 'corrected' by the Victorians and nor were vision 'errors' solely seen as pathologies. Just as spectacles were not simply tools for the 'correction' of ocular 'defects', neither were they a medical object operating in a vacuum. As has been acknowledged in broader studies of technologies and the body, medical authority can be challenged through customer autonomy and adaptations to medical devices.[58] Claire L. Jones has identified a lack of scholarly attention to the relationship between the market for assistive devices and prostheses and the conception of disability.[59] The dispensing of spectacles in a variety of sites, and often outside medical control, affected how both spectacles, and the conditions they enhanced, were perceived. Spectacles uniquely inform the ways in which medical instruments and appliances were influenced by cultural values because, unlike many other assistive devices or prosthetics, their marketing and dispensing were not set apart from mainstream culture.[60] Indeed, how technologies are used in practice is an important element of historical study; there was no one use of a device and many served and embodied multifaceted functions and meanings. Whereas, traditionally, historians and sociologists have tended to focus their analysis on the technology itself, more recent work is increasingly placing the focus on users and how technologies were disseminated and adopted more widely.[61] Teasing out popular understanding and opinion and the multifarious contexts of spectacle use is an important part of this study. A straightforward reading of the medicalisation and adoption of spectacles as an assistive device would suggest that spectacles subsequently became a device associated with the treatment of 'abnormality'. Moreover, unlike conventional prosthetics replacing or masking visible difference, spectacles emphasised a person's inability to conform to the functional norm. In certain respects, this was true but, while medical ideas could be influential, Victorian concepts of beauty and fashion were just as important in determining a person's reluctance. Exploring these narratives of resistance and non-use is important for understanding the relative value or importance of vision to the individual. For example, why might a person choose to not wear spectacles, especially if they were unable to see; was vision more important, or appearance? In contrast, a person choosing to wear spectacles could highlight the perceived value of vision and spectacles as a functional device, one that was able to overcome concerns over appearance or visible physical difference. Chris Otter postulates that the value of spectacles would have increased in tandem with the unprecedented demands and value being placed on vision by the Victorians. By illustrating these cases, in many ways this book serves to reinforce this assessment.[62]

The sites of production and dispensing, as well as design, influenced the extensive use of spectacles and complicated the parameters of the terms 'normal' and 'abnormal'. While assistive devices are often accessed through prescribed medical channels, the position of spectacles outside this medical framework is key for shaping Victorian understanding of visual impairment and visual acuity. Here, the design and distribution of assistive technology becomes important. Elizabeth Guffey and Bess Williamson argue in their design model of disability that objects and devices, and the material and digital environment, shape the meaning of disability. Their work challenges the social and medical model binary by highlighting that the design of technology independently, and profoundly, shapes contemporary understandings of 'function' and 'normality'.[63] Cara Kiernan Fallon's recent work on the walking stick, for example, reveals how a more standard, clinical design mediated the experience of walking impairments and altered the feelings associated with using a stick from pride to shame.[64] Spectacle design similarly shaped definitions of visual ability and disability. The mass-manufactured uniform frame widened accessibility and a variety of different frame designs emerged to suit users' needs: the discrete frame, the invisible frame, the bold frame, the luxurious frame, the sturdy frame and the more decorative eyeglasses. Davidson's observations were not misinformed; spectacles were becoming increasingly normalised as the demographic and number of users expanded. This is not to say that spectacles were immediately viewed as 'normal'; they embodied a range of meanings for which assistive technology increasingly catered. The very design of spectacles allowed them to perform as a multifaceted device that could convey positive social markers such as intelligence, sagacity or wealth, depending on the wearer or the context in which the device was worn. Nelly Oudshoorn and Trevor Pinch have argued that assimilation does not occur when the prescribed users or symbolic meanings attached to a piece of technology do not correspond with contemporary cultural values.[65] In contrast, assimilation can occur when these variables align. Spectacle design and their position in the mainstream market allowed them to be conceived, on one hand, as a restorative medical object and, on the other hand, a marker, even fashionable, item of display. The study of spectacles therefore reveals something about the conditions in which an assistive device can be assimilated and, therefore, the role and design of assistive technology in simultaneously creating and challenging the categorisation of normalcy and the experience of disability. Victorians initiated the popularisation of spectacles, and this was based on both the perceived importance of vision within culture and the growing acceptance and recognition of spectacles as a functional, enhancing, even in some cases 'normal', device.

Using the 'mundane' object: methodology and sources

In her overview of 'disability things' Katherine Ott argued that 'just as disability is relevant to all aspects of history its material culture is found everywhere'.[66] She concluded that the experience of disability is both shaped and mediated through objects and the environment. Guffey and Williamson, in outlining their design model of disability, similarly draw together a collection of authors that explore the historical experience of disability through artefacts. The work in *Making Disability Modern*, and Ott's chapter in that book, focuses on a variety of mundane, everyday things: the walking stick, adapted chairs, clothing, toys and an apron. Such an approach differs markedly to traditional collecting styles that have influenced the way in which the history of spectacles and visual aids has been preserved. Twentieth-century publications and collectors' findings were often informed by their interest in overall value and rarity as well as a close focus on the aid itself, for example key designs and the object's association with notable manufacturers, owners, events and/or provenance. These collecting practices were reflective of a wider twentieth-century academic tradition, which focused on luxury objects or 'high' culture.[67] A certain cross-section of objects – the valuable, significant and ornate – were prioritised over more utilitarian forms of evidence, which were often disregarded. Such an approach is markedly different to the one I present in this book. The cultural turn in the second half of the twentieth century has gradually shifted historians' focus towards the 'everyday'. The use of the term everyday does not necessarily mean 'low' or common culture.[68] Rather, it is a focus on objects or practices that form part of daily life, which include both the gold decorative eyeglass and the workaday steel spectacles. My questioning of the object also goes deeper: why were they collected? Why and how were they worn? Can the contexts and processes of wear, and changing wear, be explained?

Historians have highlighted the unique value of material evidence for illuminating the daily practices of ordinary individuals and groups of people who otherwise leave very little trace. Here, the unremarkable, plain, even broken, take central stage.[69] Material culture is therefore essential for fully exploring the historical experience of spectacle use. Research by collectors that have made more effort to consult additional forms of evidence, or have considered the types of users and functions of the frame, has shown the potential of objects – when used alongside other forms of evidence – to reveal the everyday experience of spectacle wear.[70] This book is more ambitious in scope and draws upon *circa*. 1,500 objects from the Science Museum's collections in South Kensington, London, to analyse the dispensing, design and use of Victorian everyday spectacle wear. Although the collections are largely anonymous, their very collection and materiality can answer a broad variety of historical questions. A record of consumption patterns, changes in design, patterns of associated expertise and a users' wearing experience can be found in the material form of assistive technologies and prostheses.[71] The tactile and emotive experience of handling objects, too, serves to reinforce the importance of touch in offering new ways to engage with, and research, the past.[72] Exploring the experience of users through touch offers unique insights into agency and control – particularly for marginalised groups – by allowing the historian to explore individuals' emotional attachment or how they adapted and customised their own device. Handling a spectacle frame, for example, raises a variety of questions that place the former user at the centre: who owned this frame? How did they wear it? Why did they alter it? Where did they buy it from? As museums have become more committed to documenting and displaying patient stories the materiality of the unnamed, everyday and personal has come to the fore.[73]

Incorporating a vast collection of everyday objects is not without its challenges. The utilitarian nature of most spectacles in the collection has meant that the cataloguing is often incomplete and nothing, or very little, is known about their provenance, acquisition and use. Conducting this research, therefore, was not a simple, one-way process of identifying relevant objects and then going away to undertake further research. Approaching the Science Museum's collections was a three-year, iterative interchange between object, similar objects in other collections and the development of contextual understanding through alternative, often written, sources. Anonymous collections present a very specific obstacle that lends itself to this style of working: there is no obviously related catalogue information upon acquisition. In the absence of this kind of information upon which we typically form a reliance, the materiality of the objects themselves initiated this study. In my case, the standalone objects offered opportunities to explore material, evidence of use, weight, style and inscriptions, as well as similarities and differences across frame design, to assess trends over time. Anonymous collections also encourage us to more forcibly address the complexities of how and why objects are collected. Museum objects cannot be viewed as unmediated historical records of the past and assessing the representativeness of a more miscellaneous or anonymous collection, while laborious and complex, reveals something about the psychology behind collecting. Why is this information missing? Did they care enough to acquire or keep hold of it but not to properly catalogue it? What objects are omitted in these styles of collections and what can this tell us? Answering these questions provides a window through which to view the values that lie behind how or why something was or continues to be preserved.

To mitigate problems of representativeness and better understand potential answers to these questions, this research is primarily based on three years of work with two collections amassed by Henry Wellcome and Matthew Dunscombe that now reside in the Science Museum's object stores. The collectors' contrasting collecting styles influenced the objects that they obtained and how they measured an object's relative importance. Wellcome was interested in the everyday and spectacles were mostly acquired as part of larger auction lots of various, miscellaneous, different items.[74] In contrast, Dunscombe collected more calculatedly and, typical of other collectors' approaches, focused on incorporating key examples in design or material. Dunscombe himself was a Victorian optician and his collecting interests and expertise supplement the depth and spread of Wellcome's larger collection.[75] The objects that survive therefore reveal a multifarious spectacle market that is both reflected by, and a symptom of, the collecting styles of the two men that orchestrated their collections. A study of Victorian visual aids thus cannot rely on objects alone. Such an approach, for example, would miss designs or styles that have not survived or been considered a priority by collectors, and also styles of frame and lens that were designed but never produced.

Drawing upon alternative sources and understanding the context of the objects and their collections has been a vital part of this work. Karen Harvey in her introduction to *Material Culture*, for example, defined 'material culture' against the study of 'object' or 'artefact' because this interrogates the materiality of an object and the range of contexts in which it acquires meaning.[76] The role of both text and objects in the history of medicine and science is also evidenced in recent scholarly work. Notably, Lorraine Daston has argued that objects 'talk' and do not simply 'repeat'.[77] Adam Mosley has highlighted that objects and images can complete our historical understanding rather than reinforce it.[78] Objects and textual sources offer the historian a different perspective and text can, in fact, deepen our understanding of an object's materiality. As Guffey and Williamson argue, the choices made by active agents in the design and style of an object – its producers and its users – are 'deeply contextual'.[79] Ott, similarly, emphasises the importance of exploring the choices or 'desires' associated with an object, and the way it is 'exchanged'.[80] A variety of textual sources in this book complement the Science Museum's collections by detailing how spectacles were exchanged between producer, dispenser and user, and the choices or perceptions that influenced their decisions. Far from perceiving spectacles as a 'thing', I argue that they are an assistive device influenced by, and capable of conveying, the economic, social and cultural contexts in which they were designed, sold and used.[81]

There is no single textual source base, or archive, for the study of Victorian visual aids. The history of visual aids, in part determined by the ubiquitous and utilitarian nature of spectacles, needs to be pieced together carefully across an eclectic and diverse source base.

Such a study has benefited from the digitisation of the British press, which has allowed the prevalence of spectacles and concerns about vision in Victorian Britain to be explored more thoroughly than was previously possible. Briggs' 1998 work on Victorian spectacles, for example, discussed a single article.[82] In contrast, I have considered more than 6,000 articles and just shy of 4,000 advertisements published in 44 different newspaper titles and 37 different periodical titles, catering to readership of all classes both in and outside London.[83] A number of medical texts, optical texts and medical trade catalogues have also been used, in both digital and physical form. Coupled with *The Optician* and the *British Medical Journal*, these have allowed contemporary understandings and theories of different kinds of professional to be assessed. The journals have provided information on current designs, the public use or abuse of visual aids and vision testing, and the tensions or debates surrounding professional control and professional boundaries. The Science Museum's library was an invaluable source for patents and a wide range of trade literature, including trade cards and directories. Recent studies have used patent specifications to explore the design of artificial limbs and assistive hearing devices.[84] Patents have been used similarly in this study to explore the design of visual aids, especially of styles that never reached production and are omitted from object collections. Finally, a diverse number of business archives and personal correspondence have been consulted through archival research in several locations: Cambridge, Carlisle, Chichester, Leeds, London, Nottingham, Sheffield and Somerset. The range of business archives and correspondence allowed the retail, marketing, distribution and use of visual aids to be researched, as well as ensuring that the findings were not unduly London-centric.

Exploring the objects in tandem with a wealth of historical material demonstrates that Wellcome and Dunscombe amassed, and have left, a crucial material insight into the Victorian spectacle market and spectacle wear. The key designs and luxurious frames that left a written trace in patents, contemporary advertising and portraiture are well represented but so too are the experiences of spectacle wear that have otherwise left only a faint echo in a passing line at the bottom of an advertisement referring to the 'working man's spectacles', a brief diary entry on someone's personal reflections, or in the moans of medical practitioners about itinerant quacks selling poor-quality devices. It demonstrates the value of the everyday object as a source, one that can highlight the use of material culture for exploring the experience of use or of users that otherwise leave very little trace. Through these collections I have thus been able to map evolutions in design, wearing experiences, retailing practices and stylistic choices of a far greater cross-section of spectacle users.[85] Communicating the multifarious role of the object in the formulation of these arguments has required some creativity. As scholars have noted, there are limitations in attempting to convey an object and its related information through language.[86] I have adopted two approaches based on the way that I have included material culture in this book; these, at first glance, may appear uneven across the chapters. I have included images or detailed descriptions of specific types of object where the design or materiality of that object has influenced my conclusion and driven my argument. In places this could include close object-analysis into, for example, an object's, or group of objects', weight. Elsewhere, I have attempted to replicate my experience of spending three years in these collections and the subsequent links and intimacy of knowledge I developed. Often the relationship between objects within and across the two collections more subtly prompted lines of historical enquiry or offered insight into the way the object was made, retailed, sold and used. I therefore endeavour to transport you as the reader to the rooms that housed the collections, or to some of their drawers and cupboards, to offer insight into the broader, experiential ways that material culture can inform our understanding of the past.

Chapter outline

The investigation of the sites of vision measurement, spectacle use and challenges to medical authority and/or medical definitions of normalcy is chronological. In Chapter One I examine how early Victorian understandings of vision and spectacles were divorced from medical knowledge or training and sites of medical treatment. To demonstrate this, the chapter draws upon a range of material, including medical texts, opticians' texts, material culture in the Science Museum's collections, trade cards and retailers' advertisements. It analyses the prominent role of the optician in dispensing visual aids and highlights that the scientific instrument trade was the primary site for dispensing spectacles. As part of this discussion, it argues that, first, opticians and retailers paid little attention to medical understandings of the eye and that spectacles and the enhancement of vision were marketed in the context of scientific 'accuracy' and the overall quality of the glass lenses. Second, it proposes that medical practitioners often dismissed the use of spectacles and focused on therapeutic treatments to remedy vision because of their training and concerns surrounding professionalism and quality of practice in the emerging speciality of ophthalmology. Chapter Two tracks a decisive change in the indifferent attitudes towards spectacles use within medical practice from the 1850s. It demonstrates how medical practitioners' utilisation and adaption of diagnostic technology – namely the ophthalmoscope, invented in 1851 – recategorised the meaning and understanding of blindness and medical involvement in spectacle dispensing. This chapter is fundamental to the book's overall argument. By drawing upon medical texts, medical journals and medical trade catalogues, it explores the role of medical practitioners in attempting to define both the theory related to vision enhancement and the methods of spectacle dispensing. It first explores the role of technology in the measurement of vision and defining the 'normal' eye and argues that this was decisive in transforming medical opinion on the overall utility of spectacles. Practitioners could observe the refractive condition of a person's eye for the first time and therefore could apply the principles of optics to use spectacles as a viable treatment option. Ophthalmologists, I suggest, increasingly argued that objective methods needed to be used in the dispensing of spectacles and that these should be conducted only by those who were medically qualified. By cultivating this discrete body of expertise, medical practitioners attempted to leverage the importance of vision and created new sites to test vision and dispense spectacles, to help justify their increasing intervention. I argue that ophthalmologists increasingly advocated and were involved in testing vision in both schools and the workplace.

Chapters Three and Four assess the limits of medical practitioners' attempts to cultivate and promote their area of expertise between 1850 and 1900. In Chapter Three I demonstrate that the expansion of spectacle manufacture and changing features of spectacle design had a decisive influence on medical practitioners' inability to assert control. As a result of mass manufacture and a more uniform, well-fitting frame, the accessibility and functionality of spectacles reached unprecedented heights. The chapter investigates how the scale of spectacle manufacture and dispensing was an obstacle in medical attempts to encroach on and monopolise vision testing and spectacle dispensing. This, I suggest, complicates medical definitions of normalcy, abnormalcy and disease and therefore challenges the parameters devised by medical practitioners in the 1850s. This chapter draws heavily upon material culture and explores the specificities of design to highlight how manufacture developed and how certain design features – particularly the side-arms of spectacle frames – enhanced their overall usability. I argue first that medical definitions of 'normalcy' and 'abnormalcy' were too simplistic and failed to capture the diversity of visual capacity in Victorian Britain; and second, that the scale of spectacle manufacture affected medical practitioners' ability to claim and exercise their area of expertise. Here, I tease out some of the peculiarities of visual aids, and their adaptability for mass manufacture, in comparison to other forms of assistive device and prosthesis. In Chapter Four I then shine light on how the scope of spectacle production expanded the spectacle market outside medical control. It highlights the subsequent tensions and shared concerns that emerged between opticians and medical men about professional jurisdiction over an increasingly lucrative market. It draws upon material culture and a broad and extensive range of archival and digital sources: advertisements, medical texts, medical journals, *The Optician*, opticians' texts, newspapers and periodicals. I argue that the 1890s were an intense period of inter- and intra-professional debate between ophthalmologists and opticians. In exploring popular responses to sight loss, I demonstrate that, while opticians were effective in maintaining their position as experts, both opticians and medical practitioners' authority as experts was challenged by the increasing availability of spectacles from miscellaneous high street retailers and conflicting popular advice. This, I suggest, demonstrates the uniqueness of spectacles as a common and ubiquitous assistive device; there was a divide in popular opinion, and many believed that the dispenser did not need to possess professional expertise. Opticians and ophthalmologists increasingly collaborated to regulate dispensing practices against a backlash and popular demand for cheap or highend frames on the high street that paid little consideration to their functionality.

Chapter Five shifts the book's focus to the experience of wearing Victorian visual aids. It covers the breadth of this book's period from 1830 to 1904, to investigate how spectacles and eyeglasses refashioned the meanings of blindness and the ways in which their wear mediated attitudes to partial sight and the use of an assistive technology. It asks three questions: to what extent was Victorian vision transformed? Who wore this technology? Why was there a diverse range of social and cultural meanings? The chapter draws upon popular literature in newspapers and periodicals, opticians' account books, medical case accounts, material culture and photographs. In this final chapter, I highlight that, like some other accessories of dress such as the Victorian fan and parasol, visual aids were not a trivial addition but instead performed subtle ideological work.[87] To wear visual aids was to fashion the face, which encompassed far more than the simple appearance of a frame; spectacles and eyeglasses shaped and responded to a variety of attitudes associated with their use. Visual aids could be actively sought, even by those who did not need them, to portray, for example, intelligence or wealth; or be avoided by those who did need them but were more concerned about stigmatic associations with age, disease and/or defectiveness. By exploring the elderly wearer, the self-conscious young woman fashioning her dress, the young man displaying eyewear or attempting to obtain work, the educated man or woman adopting a visual aid to bolster their reputation, or those seeking to hide facial difference, I demonstrate that perceptions about, and responses to, the use of visual aids were deeply affected by the gender, age and class of the wearer. I suggest that popular responses to visual aids both reinforced and challenged the medical definitions of normalcy and abnormalcy and the scope for medical practitioners to medicalise or monopolise the dispensing of spectacles; spectacles were not simply a medical assistive device and instead embodied a more diverse range of functions and meanings that were shaped by popular opinion. The adaptability of visual aid frame design to serve the needs of its users was vital in enhancing its popularity and shaping contemporary understanding of partial sight.

The conclusion draws together the thread of technology and medical intervention which has been woven into each chapter of the book. It examines how technology mediated the Victorian experience of seeing, both by determining the parameters of being able to 'see' and by offering an opportunity to subvert medical categories and medical control. I look forward and demonstrate how the blurred boundaries between different types of expertise – medical, optical and the experiential gained by spectacle wearers – influenced twentieth-century spectacle dispensing practices and styles up until the Optician's Act in 1958, and beyond. I also argue that my approach to vision measurement – and its shaping of bodily capacity – demonstrates a real opportunity to connect recent work in sensory history to growing trends in disability history, which both seek to explain sensory capacity as a varied, deeply contextual and subtle continuum. Assistive technologies offer a unique case study for exploring how technology, and the cultural or physical environment in which it is used, influence the ways bodily capacity is understood and responded to. The rapid expansion, even 'maturation', of industry, urbanisation and medical investigation experienced by the Victorians offers an ideal case study to initially begin this work.[88] The utility of assistive technology, medical knowledge and the degree to which it was accepted or undermined, determined the levels to which certain capabilities were normalised and others, by the Victorian period, were sub-standard and demanded enhancement.

Notes

1 Vincent Ilardi, *Renaissance Vision from Spectacles to Telescopes* (Philadelphia, PA: American Philosophical Society, 2007), pp. 75–79, 128.

2 Alun Withey, *Technology, Self-Fashioning and Politeness in Eighteenth-century Britain: Refined Bodies* (Basingstoke: Palgrave Macmillan, 2016), pp. 97–98.
3 Ibid., pp. 91–112.
4 Dr Dyce Davidson, 'Professor Davidson on the eyesight: the evils of our school system', *Aberdeen Journal* (18 February 1885).
5 For an overview see, for example, *Ophthalmic Antiques Collectors Club Bulletin* (East Chillington: The Club, 1982–1985) and *The Newsletter: Ophthalmic Antiques International Collectors Club* (East Chillington: The Club, 1985 to present), available at the Science Museum's Library or from the Club.
6 Withey, *Self-Fashioning and Politeness*, pp. 91–112; Kerry Segrave, *Vision Aids in America: A Social History of Eyewear and Sight Correction since 1900* (London: McFarland & Company Inc., 2011); Neil Handley, *Cult Eyewear: The World's Enduring Classics* (London: Merrell Publishers Ltd, 2011); Asa Briggs, *Victorian Things* (London: B.T. Batsford Ltd, 1998), pp. 103–141.
7 Briggs, *Victorian Things*.
8 See, for example, Martin Jay, 'The rise of hermeneutics and the crisis of ocularcentrism', *Poetics Today*, 9:2 (1988), 307–326 and reflections on recent scholarship by Mark M. Smith, *A Sensory History Manifesto* (Philadelphia: Pennsylvania State University Press, 2021).
9 Graeme Gooday and Karen Sayer, *Managing the Experiences of Hearing Loss in Britain, 1830–1930* (Basingstoke: Palgrave Macmillan, 2017), pp. 48–49.
10 See, for example, the work of Gooday and Sayer, Jaipreet Virdi and Coreen McGuire referenced throughout this book.
11 Christopher Michael Woolgar, *The Senses in Late Medieval England* (New Haven, CT; London: Yale University Press, 2006), p. 1.
12 Chris Otter, *The Victorian Eye: A Political History of Light and Vision in Britain, 1800–1910* (Chicago, IL: University of Chicago Press, 2008), pp. 25–28; Jonathan Crary, *Techniques of the Observer: On Vision and Modernity in the Nineteenth Century* (Cambridge, MA: MIT Press, 1990).
13 Otter, *The Victorian Eye*, pp. 54, 61; Rosemarie Garland-Thomson, *Staring: How We Look* (Oxford: Oxford University Press, 2009), p. 26; Kate Flint, *The Victorians and the Visual Imagination* (Cambridge: Cambridge University Press, 2000), p. 8; Peter John Brownlee, *The Commerce of Vision: Optical Culture and Perception in Antebellum America* (Philadelphia: University of Pennsylvania Press, 2019).

14 Smith, *A Sensory History Manifesto*, pp. 29–31, 79; see also Jonathan Reinarz, *Past Scents: Historical Perspectives on Smell* (Urbana: University of Illinois Press, 2014); Alain Corbin, *Time Design and Horror: Toward a History of the Senses*, trans. by Jean Birrell (Cambridge: Polity Press, 1995); Alain Corbin, *The Foul and The Fragrant: Odor and the French Social Imagination* (Cambridge, MA: Harvard University Press, 1996).
15 Smith, *A Sensory History Manifesto*, p. 65.
16 David M. Turner, 'Disability history and the history of emotions: reflections on eighteenth-century Britain', *Asclepio*, 68:2 (2016). DOI: 10.3989/asclepio.2016.18.
17 Gooday and Sayer, *Managing the Experiences of Hearing Loss*, p. 49.
18 Gordon Phillips, *The Blind in British Society: Charity, State and Community c.1780–1930* (Aldershot: Ashgate, 2004); Matthias Reiss, *Blind Workers against Charity: the National League of the Blind of Great Britain and Ireland, 1893–1970* (Basingstoke: Palgrave Macmillan, 2015); Fred Reid, 'The Panopticon: towards an intimate history of special schools for the blind', in Iain Hutchison, Martin Atherton and Jaipreet Virdi (eds), *Disability and the Victorians: Attitudes, Interventions, Legacies* (Manchester: Manchester University Press, 2020), pp. 164–176.
19 'Some social changes in fifty years', *The Nineteenth Century: A Monthly Review* (March 1892), p. 465.
20 *Daily News* (13 September 1890).
21 See, for example, Georg Simnel and Christoph Asendorph quoted in Otter, *The Victorian Eye*, pp. 23–24.
22 Brownlee, *The Commerce of Vision*, p. 19; see also, Otter, *The Victorian Eye*, p. 40.
23 See, for example, Simeon Snell, *Influences of School Life on Eyesight* (London: Wyman & Sons, 1884), p. 6; 'The eyesight of school children', *Aberdeen Weekly Journal* (27 January 1890); 'Scientific and industrial notes', *Manchester Times* (26 December 1890); 'Miscellaneous news', *The Belfast News-Letter* (30 December 1890); 'Jottings: miscellaneous news', *The Star* (30 December 1890); 'Miscellaneous news', *The Yorkshire Herald* (30 December 1890).
24 See, for example, Brownlee, *The Commerce of Vision*, pp. 11–12.
25 See, for example, George Weisz, *Divide and Conquer: A Comparative History of Medical Specialisation* (New York: Oxford University Press, 2006); Lindsay Granshaw, '"Fame and fortune by means of bricks and mortar": the medical profession and specialist hospitals in Britain, 1800–1948', in Lindsay Granshaw and Roy Porter (eds),

The Hospital in History (London: Routledge, 1989), pp. 199–220; George Rosen, *The Specialisation of Medicine with Particular Reference to Ophthalmology* (New York: Froben Press, 1944).

26 For ophthalmology see, for example, Luke Davidson, '"Identities ascertained": British ophthalmology in the first half of the nineteenth century', *Social History of Medicine*, 9:3 (1996), 313–333.

27 An exception would be Diane D. Edwards, 'Optometry', in Daniel M. Albert and Diane D. Edwards (eds), *The History of Ophthalmology* (Oxford: Blackwell Science, 1996), pp. 303–310.

28 Wolfgang H. Vogel and Andreas Berke, *Brief History of Vision and Ocular Medicine* (Amsterdam: Wayenborgh Publishers, 2009), pp. 221–223; Mary Carpenter, *Health, Medicine and Society in Victorian England* (Santa Barbara, CA: Praeger, 2009), pp. 143–144; Otter, *The Victorian Eye*, p. 38.

29 See, for example, Carpenter, *Health, Medicine and Society*, pp. 12–13; Stanley Joel Reiser, *Medicine and the Reign of Technology* (Cambridge: Cambridge University Press, 1978).

30 See, for example, Roy Porter, *Health for Sale: Quackery in England, 1660–1850* (Manchester: Manchester University Press, 1989), pp. 222–235; Irvine Loudon, 'Medical practitioners 1750–1850 and the period of medical reform in Britain', in Andrew Wear (ed.), *Medicine in Society* (Cambridge: Cambridge University Press, 1992), pp. 219–247; Christopher Lawrence, *Medicine in the Making of Modern Britain, 1790–1920* (London: Routledge, 1994); Keir Waddington, 'Mayhem and medical students: image, conduct, and control in the Victorian and Edwardian London teaching hospital', *Social History of Medicine*, 15:1 (2002), 45–64; Michael Brown, 'Medicine, reform and the "end" of charity in early nineteenth-century England', *English Historical Review*, 124:511 (2009), 1353–1388; Michael Brown, *Performing Medicine: Medical Culture and Identity in Provincial England, c.1760–1850* (Manchester: Manchester University Press, 2010).

31 Michel Foucault, *The Birth of the Clinic: An Archaeology of Medical Perception* (New York: Vintage Books, 1975), pp. 33–34.

32 For an overview see Peter Conrad, *The Medicalisation of Society: On the Transformations of Human Conditions into Treatable Disorders* (Baltimore, MD: The Johns Hopkins University Press, 2007), pp. 3–4.

33 Michel Foucault, *The History of Sexuality*, vol. 1 (London: Penguin Books, 1990), p. 44.

34 Carpenter, *Health, Medicine and Society*, p. 144.

35 For more information on this see, Gemma Almond, 'Vision testing in late nineteenth and early twentieth-century Britain: opticians, medical

practitioners and the battle for professional authority', *Social History of Medicine*, 35.1 (2022), 237–258. DOI: 10.1093/shm/hkab122.
36 Takahiro Ueyama, *Health in the Marketplace: Professionalism, Therapeutic Desires, and Medical Commodification in Late-Victorian London* (California: SPSS, 2010).
37 Ibid. and Claire L. Jones, *The Medical Trade Catalogue in Britain, 1870–1914* (London: Pickering and Chatto, 2013).
38 Ueyama, *Health in the Marketplace*, pp. 22, 107.
39 Claire L. Jones (ed.), *Rethinking Modern Prostheses in Anglo-American Commodity Cultures 1820–1939* (Manchester: Manchester University Press, 2017), pp. 1–23.
40 For the eighteenth century see, for example, Liliane Hilaire-Pérez and Christelle Rabier, 'Self-machinery? Steel trusses and the management of ruptures in eighteenth-century Europe', *Technology and Culture*, 54:3 (2013), 460–502 (p. 490).
41 Conrad, *The Medicalisation of Society*, p. 9; Robert A. Nye, 'The evolution of the concept of medicalization in the late twentieth century', *Journal of History of the Behavioural Sciences*, 39:2 (2003), 115–129 (p. 122).
42 See, for example, the use of differing terms in Katherine Ott's work from 'orthotic' to 'prosthetic': Katherine Ott, 'The sum of its parts: an introduction to modern histories of prosthetics', in Katherine Ott, David Serlin and Stephen Mihm (eds), *Artificial Parts and Practical Lives: Modern Histories of Prosthetics* (New York: NYU Press, 2002), p. 7; Katherine Ott, 'Prosthetics', in Rachel Adams, Benjamin Reiss and David Serlin (eds), *Keywords for Disability Studies* (New York: New York University Press, 2015), p. 140.
43 Ott, 'Prosthetics', p. 140.
44 Chapal Khasnabis, Zafar Mirza and Malcolm MacLachlan, 'Opening the gate to inclusion for people with disabilities', *The Lancet*, 386:10010 (2015), 2229–2230.
45 Ott, 'The sum of its parts', p. 7.
46 D.A. Caeton, 'Blindness', in Adams, Reiss and Serlin (eds), *Keywords for Disability Studies*, p. 35.
47 Jaipreet Virdi and Coreen McGuire, 'Phyllis M. Tookey Kerridge and the science of audiometric standardisation in Britain', *British Journal for the History of Science*, 51:1 (2018), 123–146.
48 Katherine Ott, 'Disability things: material culture and American disability history, 1700–2010', in Susan Birch and Michael Rembis (eds), *Disability Histories* (Chicago: University of Illinois Press, 2010), p. 121.

49 Definition of disability under the Equality Act 2010 – GOV.UK (archive.org) (accessed: 23 April 2022).
50 Lennard J. Davis, *Enforcing Normalcy: Disability, Deafness, and the Body* (London: Verso, 1995), pp. 24–26.
51 Coreen McGuire, *Measuring Difference, Numbering Normal: Setting the Standards for Disability in the Interwar Period* (Manchester: Manchester University Press, 2020).
52 F.C. Donders, *On the Anomalies of Accommodation and Refraction of the Eye, with a Preliminary Essay on Physiological Dioptrics*, trans. William Daniel Moore (London: New Sydenham Society, 1864).
53 Peter Cryle and Elizabeth Stephens, *Normality: A Critical Genealogy* (Chicago, IL: University of Chicago Press, 2017).
54 Ibid., p. 3.
55 For a parallel argument see Virdi and McGuire, 'Phyllis M. Tookey Kerridge'.
56 Gooday and Sayer, *Managing the Experiences of Hearing Loss*, pp. 2–3.
57 McGuire, *Measuring Difference*; see also Jaipreet Virdi, 'Medicalising deafness in Victorian London: The Royal Ear Hospital, 1816–1900', in Hutchison, Atherton and Virdi (eds), *Disability and the Victorians*, pp. 73–91.
58 Bess Williamson, 'Electric moms and quad drivers: people with disabilities buying, making, and using technology in postwar America', *American Studies*, 52:1 (2012), 5–30; Nelly Oudshoorn and Trevor Pinch (eds), *How Users Matter: The Co-Construction of Users and Technology* (Cambridge, MA: MIT Press, 2005); Ruth Oldenziel and Mikael Hard, *Consumers, Tinkerers, Rebels: The People Who Shaped Europe* (Basingstoke: Palgrave Macmillan, 2013).
59 Jones, *Rethinking Modern Prostheses*, p. 2.
60 In contrast see, for example, Williamson, 'Electric moms and quad drivers', p. 11.
61 Oudshoorn and Pinch (eds), *How Users Matter*, p. 12.
62 Otter, *The Victorian Eye*, pp. 39–40.
63 Elizabeth Guffey and Bess Williamson (eds), *Making Disability Modern: Design Histories* (London: Bloomsbury, 2020), pp. 1–13.
64 Cara Kiernan Fallon, 'Walking cane style and medicalised mobility' in ibid., pp. 43–60.
65 Oudshoorn and Pinch (eds), *How Users Matter*, p. 19.
66 Ott, 'Disability things', p. 129.
67 Tara Hamling and Catherine Richardson (eds), *Everyday Objects: Medical and Early Modern Material Culture and its Meanings* (Farnham: Ashgate Publishing, 2010), p. 13.

68 Ibid., pp. 13–14.
69 Sara Pennel, 'Mundane materiality: or should small things still be forgotten?' in Karen Harvey (ed.), *History and Material Culture* (Abingdon: Routledge, 2009), pp. 173–191.
70 See, for example, Neil Handley's study of prominent brands in *Cult Eyewear*; and numerous short articles in the *Ophthalmic Antiques Collectors Club Bulletin* and *The Newsletter: Ophthalmic Antiques International Collectors Club*.
71 Ott, Serlin and Mihm (eds), *Artificial Parts and Practical Lives*, throughout; Neil Handley, 'Artificial eyes and the artificialisation of the human face', in Carsten Timmermann and Julie Anderson (eds), *Devices and Designs: Medical Technologies in Historical Perspectives* (Basingstoke: Palgrave Macmillan, 2006), pp. 97–111.
72 Ott, 'Disability things', pp. 120, 133; Emily Robinson, 'Touching the void: affective history and the impossible', *Rethinking History: The Journal of Theory and Practice*, 14:4 (2010), 503–520 (p. 503).
73 Sophie Goggins, Tacye Phillipson and Samuel J. M.M. Alberti, 'Prosthetic limbs on display: from maker to user', *Science Museum Group Journal*, 8 (2017). DOI: 10.15180/170806 (accessed 28 March 2019).
74 The surviving auction catalogues marked up by his collectors, for example, document the acquisition of visual aids for a low price and amongst a vast array of miscellaneous items. These are currently uncatalogued but are available at the Wellcome Library. Catalogues for Stevens' auction house were viewed for 2 March 1915; 13 April 1920; 2 September 1924; 21 and 22 August 1928; 21 September 1928; 14 October 1928; 23 October 1928; 26 and 27 March 1930; 16 and 17 September 1930; 7 November 1930; 5 August 1931; 20 and 21 October 1931; 20 and 21 September 1921; 20 and 21 October 1932; 16 April 1935; and catalogues for Glendining's auction house for 29 July 1932; 29 October 1934; 14 January 1935. See also Ghislaine M. Skinner, 'Sir Henry Wellcome's Museum for the Science of History', *Medical History*, 30 (1986), 383–418; Frances Larson, 'The things about Henry Wellcome', *Journal of Material Culture*, 15 (2010), 83–104; Ken Arnold and Danielle Olsen (eds), *Medicine Man: The Forgotten Museum of Henry Wellcome* (London: The British Museum Press, 2011), pp. 29–50.
75 Science Museum Technical File, London, T/1921-323, Catherine Gates, 'Matthew William Dunscombe: a Bristol optician', research paper (1997), Derek C. Davidson, 'Matthew William Dunscombe the first great collector of antique spectacles', *Ophthalmic Antiques Newsletter*, 4, p. 51, and

Margaret Mitchell, 'Optics and the Science Museum', *The Optician* (21 September 1979), pp. 22–24.
76 Harvey (ed.), *Material Culture*, p. 3.
77 Lorraine Daston (ed.), *Things that Talk: Object Lessons from Art and Science* (New York: Zone Books, 2004), p. 11.
78 Adam Mosley, 'Introduction [to special issue "Objects, texts and images in the history of science"]', *Studies in the History & Philosophy of Science*, 38 (2007), 289–301 (p. 291).
79 Guffey and Williamson, 'Introduction' in *Making Disability Modern*, p. 8.
80 Ott, 'Disability things', p. 122.
81 Jones and Withey have made a similar argument that prosthetics are not just 'things': see Jones, *Rethinking Modern Prostheses*, p. 5; Withey, *Self-Fashioning and Politeness*, throughout.
82 Briggs, *Victorian Things*, p. 105.
83 A variety of terms were used, to overcome problems that emerge from either the character recognition technology or the historical use of alternative terminologies. These include 'eyeglasses', 'spectacles', 'glasses', 'glasses and lenses', 'optician', 'oculist', 'hypermetropia', 'hypermetropic', 'myopia' and 'myopic'. I consulted 2,115 newspaper articles, items of correspondence and reports in sports and news sections, and over 1,000 newspaper advertisements. I also consulted 4,089 articles, reviews, 'general information' and correspondence, as well as 2,818 advertisements in Victorian periodicals. The position as well as the content of the key word search result was analysed to ensure that it was not removed from its context or format on the page.
84 Jones, *Rethinking Modern Prostheses*, p. 17.
85 Findings from these collections were cross-referenced with the collection at the College of Optometrists, to assess how typical they were.
86 Ott, 'Disability things', p. 123.
87 Susan Hiner, *Accessories to Modernity: Fashion and the Feminine in Nineteenth-century France* (Philadelphia: University of Pennsylvania Press, 2010); Ariel Beaujot, *Victorian Fashion Accessories* (London: Bloomsbury, 2012).
88 Hutchison, Atherton and Virdi (eds), *Disability and the Victorians*, p. 14.

1

Early Victorian understandings of vision and spectacles, 1830–1850

During 1840 a London optician, John Thomas Hudson, distributed a pamphlet on spectacles. In many ways, there are parallels between Hudson's observations and those made in 1885 by Dr Dyce Davidson, quoted in the Introduction. Hudson observed that the number of spectacle wearers was growing during these first years of Victoria's reign. He argued that:

> [t]he use of Spectacles by persons of all ages has of late years become so prevalent – the advantages derived from their use so evident – the commerce produced by their manufacture so extensive – and the assistance rendered by the Optician so great, that no apology will be required from the writer in thus offering to the public, ample information relating to Spectacles and Spectacle Lenses.[1]

Unlike Davidson, Hudson did not place the medical practitioner at the centre of his story of progress; rather, he focused on his own profession: the optician. In Hudson's day, the optician was the figure long associated with spectacles and spectacle dispensing. This is perhaps not surprising to today's reader, considering the optician's continued role in spectacle dispensing. Traditionally, however, the term 'optician' did not denote expertise in spectacle retail and instead signified a knowledge of applied optics and specialisation in the manufacture of optical lenses and instruments such as the telescope and microscope.

Hudson did not confine his role to what was customary. The optician, in his eyes, bolstered the commercial sale and retailing of spectacles and acted as a public expert, offering advice on spectacle use and vision testing. How accurate is Hudson's viewpoint?

Why might Hudson's views on vision testing be so startling? Where and how would the early Victorian encounter spectacles? Were medical practitioners involved? This chapter explores the locations of expertise and the intended dispensers of spectacles in the early Victorian period. Capitalising on the expansion of print advertising, spectacles, and a host of corrective body devices, emerged in eighteenth-century print to restore functionality or an image of physical wholeness.[2] Although traders in this new type of commodity sought to cultivate a relationship with the medical faculty, often these devices were not strictly medical objects.[3] To classify early Victorian spectacles as a medical object is even more doubtful. Traders in the eighteenth and early nineteenth centuries did not seek to advertise any obvious relationship with the medical faculty. Likewise, ophthalmology – the first successful medical specialty to be established in Britain – included neither vision testing nor spectacle dispensing as part of its initial disciplinary practice. At a time when medical specialisation was viewed with mistrust, ophthalmologists instead focused on developing therapeutic, observable, clinical principles. When spectacles were deemed medically required, ophthalmologists and opticians worked within clearly defined jurisdictional boundaries; it was the optician that dealt with the separate practice of vision testing and spectacle dispensing.

In this chapter I seek to demonstrate that the early Victorian did not encounter, interact with or buy spectacles in a medical setting. In the eighteenth century spectacles had been part of a broader array of corrective devices that occupied a liminal position in the trade in scientific instruments and toys. On one hand, they could be a highly specialised, bespoke item. On the other hand, they could be a quotidian 'impulse buy' that served to bulk out a retailer's stock.[4] In the early Victorian period spectacles were categorised in similar ways but retailers took advantage of the established commercial potential of the product, the growing number of retail outlets and the comparative lack of regulation.[5] The rise of increasingly specialised shop premises – for example, the scientific instrument maker, the jeweller, the watchmaker or the engraver – and the persistence and adaptation of the sundry trade operating in more miscellaneous spaces such as the street and bazaar altered the visibility of spectacle sale. Such expansion was possible because the trader did not need to possess any form of expertise and the sale of spectacles was configured as a standard commercial transaction. Miscellaneous traders did not often claim to be 'oculists' and neither did they seek to assert knowledge of the eye; this type of advertiser utilised bold typeface to direct their reader's attention to the fact they simply stocked a variety of spectacles.

The expansion in a diverse, but unregulated, retail market is a distinctive feature of the early Victorian period. I argue, like Hudson, that the reputable trader faced an unprecedented challenge. The many 'advantages' of spectacles as an aid to restoring distant or near vision were increasingly acknowledged by the public in tandem with the rising number of traders seeking to profit from their sale. Those seeking to 'protect' the nation's eyes, or gain a commercial monopoly over a growing trade, tackled the puffing exhortations of sellers in provincial and metropolitan newspapers, streets and shops by penning their own advertisements and treatises to expose wrongdoing. Contemporary understandings of the eye and how spectacles worked are a vital part of this narrative. While retailers and the general population acknowledged the vision-enhancing properties of spectacles, exactly how optical lenses enhanced vision was unknown and divorced from any observable medical understanding of the refractive capacity of an individual's eye, or its anomalies. Contemporary categories of vision – good and weak – were influential in circumventing medical intervention and the possibilities of objective vision testing methods. 'Weak vision' comprised two conditions: 'short sight', which affected a person's ability to see over long distances, and 'old sight', which affected a person's ability to read in advancing age. Optical and medical understandings of these phenomena were limited to abstract theories of lenses and optical refraction. These theories could explain *why* a person required a concave or convex lens, but they did not help adjust or calculate the gradation of the lens to the specific construction of an individual eye based on any form of observable evidence. All traders therefore had to adopt methods of trial and error, which could often lead to spectacles being perceived as a simple and unsophisticated device capable of generic application and dispensed to the population at large. Ill-conceived ideas abounded, including that the age of the wearer or the distance that they could see for conventional activities, such as reading or eating, determined the lens that was required.[6]

In the 1830s Hudson was part of a group of opticians that desired to 'correct' the public's misconceptions and present spectacles as a complex device having the potential to damage a person's vision when misapplied; his process of trial and error was informed by expertise. These retailers capitalised upon medical interest in the eye to construct an expert 'skillset' capable of contending with the putative conduct of 'quacks' penetrating the broader medical marketplace for patent medicines and technologies. Hudson was attempting, like his predecessors selling steel trusses, to establish a unique skillset that combined applied scientific knowledge with anatomical knowledge.[7] Crucially for Hudson, this was a skillset that could only be constructed outside traditional medical training and practice.

By drawing upon material culture, opticians' treatises, advertisements, accounts in popular literature, opticians' account books and medical texts, this chapter explores, first, the location and extent of the early Victorian spectacle market, and second, how a select group of traders eventually sought to introduce a more scientific style of spectacle dispensing distinct from, but in partnership with, the discipline of ophthalmology. While earlier studies have considered spectacle dispensing practices in this period to be 'haphazard' and unformalised, I argue that such a view is anachronistic; the process of trial and error, which allowed the customer to try a variety of lenses of different gradations to decide which suited their vision best, was adapted by a group of concerted opticians and ophthalmologists and laid the foundations for more objective, thorough standards for testing vision. Nevertheless, despite these efforts, spectacles' restorative properties remained shrouded in mystery. Spectacles were a lucrative, commercial device the function of which was neither understood nor realised in full. To the early Victorian spectacles were a scientific instrument or bespoke aid sold by the optician, a decorative accessory sold by the jeweller, or a quotidian, assistive device obtained cheaply from itinerant traders in the street. This very ambiguity has left a legacy crucial for understanding why medical, optical and popular responses to the wear and retail of spectacles was, and remains, so fractured.

The reputable optician and the retail of spectacles

Time spent in the Science Museum's Ophthalmology collection, with its drawers of spectacles and cupboards of vision testing equipment, unveils the sites from which spectacles were dispensed in the early Victorian period, and which encouraged the formation of a group of 'expert' retailers. The material evidence, what is present and what is missing, reveals that the early Victorians encountered spectacles in a variety of locations. By initially focusing on the retail and marketing of spectacles I am not saying that the manufacture of spectacles in this period was less significant. Metallurgical manufacturing experience and innovation influenced the trades that engaged in the retail of corrective body devices.[8] For example, traders involved in the manufacture of decorative metal products – such as jewellers or watchmakers – made and sold spectacles, as did scientific instrument makers. Moreover, the desirability of self-manufacture can be seen in advertising claims that initial manufacture or repair was undertaken on the premises.[9] However, manufacture was not a requisite for legitimising an early Victorian's ability to supply spectacles and, certainly, not all opticians, scientific instrument makers or traders in fancy goods both made and supplied them. The growing number of spectacles supplied and distributed, and then sold on, by retailers helps to explain the expansion of the spectacle market. The increasing separation between manufacture and retail had a rippling effect across the Victorian period, which will be explored in detail in Chapters Three and Four. This chapter sees opticians' claims of expertise as intriguing and complex; while, by this time, many were simply suppliers like other traders, with no manufacturing skill or experience, opticians could draw upon their traditional role in spectacle and optical lens manufacture to bolster a reputation for having a new and increasingly anatomical understanding of the eye.

The language of science appears on the early Victorian object, but not that of medicine. The trade and location of the retailer carefully inscribed on the spectacle cases and frames that they sold reveal a retailing landscape dominated by opticians and scientific instrument makers. Of these, eighty-eight could be traced to a corresponding trade directory and reveal how retailers chose to identify themselves.[10] Although the standalone optician was the largest category, it was outnumbered by a range of optical, mathematical and philosophical instrument makers, plus a more limited number of specialised 'spectacle makers' and a goldsmith. Nearly all the traders in the collection had a London address.[11] Was the sale of early Victorian spectacles concentrated in the metropolis? Did the market consist primarily of specialist retailers? How representative are these collections' findings? While inscriptions on the objects are a rich source of information for makers and suppliers, the 'named' information on objects does have its limitations. Inscriptions on scientific instruments in particular might lead the incautious to suppose that the scientific instrument trade was predominantly London-based.[12] In reality, the trade thrived in provincial centres as a result of the diversification of high street shops in a range of towns.[13] Scholars, for example, have argued that, by 1851, provincial instrument makers were confident enough to appear in the Great Exhibition and, in some instances, were able to compete against London makers.[14] Studies of spectacles, although often vague in this regard, similarly suggest that the sale of spectacles had spread into provincial centres either through expanding retail premises or the itinerant pedlar.[15]

While historians have typically focused on the earlier so-called birth of consumerism and the later effects of industrialisation, the retail of spectacles in the first half of the nineteenth century benefited from population expansion, the growth and prominence of shop premises and changes in income and expenditure.[16] It is important, however, not to overestimate these changes. Spectacles, like other areas of retail, were sold in both 'shop' and 'beyond the shop' forms of trade. Hitherto, historical studies of spectacle dispensing have paid attention to only two primary methods of distribution: via itinerant pedlars on the street and opticians' premises.[17] Previous work therefore has not fully appreciated the complexity of the early Victorian retail market. Spectacles presided in the expanding, specialised shop premises but were also offered in miscellaneous spaces of sale, including purpose-built halls such as bazaars and by pedlars on the street who thrived and helped to support the growth in consumer demand.[18] If we move away from inscriptions, the materiality of the frames and a variety of associated trade literature (including advertisements and business accounts) reveal a far more diverse and geographically wide retail market. I would argue that the anonymity of most frames – a large proportion of them carry little information about where they were made, sold or bought – is highly demonstrative of early Victorian retailing practices. The involvement of goldsmiths, jewellers, fancy-goods makers and cutlers leaves traces in the materials of the frames they sold and their appearance in a variety of early Victorian metropolitan and provincial print. Moreover, the anonymity of a vast array of more affordable steel spectacles offers an insight into a retailing practice that has otherwise left very little historical trace.

The newspaper with its dense columns of advertising had become a vehicle for retailers to market their products and commodities, especially after the reduction of the advertisement duty in 1833.[19] The expansion of adverts claiming to 'care for the eyes', or 'restore' or 'preserve' the sight through the purchase of spectacles of 'special qualities', is simultaneously both ironic and calculating. As Peter John Brownlee has demonstrated in his study of antebellum America, the printing of an advertisement for spectacles supported and profited the industry by exposing a person's deteriorating visual acuity and contributing to their eye strain.[20] In Britain, the early Victorian spectacle seller capitalised on concerns about deteriorating eyesight, a desire to preserve eyesight and a need to read small print. The optician and scientific instrument maker stood at the forefront and advertised the sale of spectacles prolifically in a range of locations. The widespread reach of opticians' advertising followed wider trends in retail, which suggest that by 1830 many towns had a diverse number of retailers and services supplying both luxury and non-luxury items.[21] Early Victorians would have observed that the optical trade was thriving.[22] While the database is not comprehensive, provincial opticians selling spectacles are scattered across the British Library Newspaper archive in a variety of towns that are part of its geographical reach: northern Ireland, north Wales, Scotland and the north and south of England. Traders also benefited from contemporary itinerant retailing practices to broaden their market range. Opticians, whether based in London or provincial centres, advertised any travel outside their localities. This could involve door-to-door service to extend their reach or to meet the needs of customers who were unable (or unwilling) to travel.[23] In 1837, a 'manufacturer of the Eye-preserving spectacles' of 37 Broad-Street, Bloomsbury, advertised in a London newspaper that he was able to attend his customers 'at their own residence within ten miles of London'.[24] Several opticians travelled further afield and would take business premises in several towns on a temporary basis. Typically, the sellers arrived in a certain area for a fixed number of days and advertised their whereabouts in the local newspaper, to call public attention to their presence. Advertisements from Messrs Davis and Sons, based in St James Street in London and 3 Mulberry Street in Liverpool, illustrate the travel an optician was capable of undertaking in this way. Between 1837 and 1840, the firm advertised in at least six different publications, on at least twelve occasions, that its representatives were available to local residences 'for a few days' or 'a short time' in places as far afield as north Wales, Bradford, Derby, Preston and Blackburn.[25] Similarly, other opticians travelled on a more regular basis to towns in their local region. In 1850, M. & N. Gluckstein, opticians of 24 Turner Street in east London, visited Ipswich every day and repeatedly advertised their presence in Bury St Edmunds.[26]

It does not take a discerning reader to notice the number of different traders advertising the sale of spectacles. Early Victorian print conjures up an image of spectacles featuring in the shop windows of a variety of premises lining the growing high streets and towns. Only one jeweller's name can be found inscribed on the case or frame of a pair of spectacles in the Science Museum's Ophthalmology collections: Benjamin West, a London 'Goldsmith and Optician' active in 1828.[27] However, the collections are home to a variety of ornate and simpler frames made from luxurious materials. Those engaged in metal work frequently advertised spectacles in a range of printed publications and could legitimately manufacture, market and sell spectacles. While an unusual combination by twenty-first-century standards, several traders styled themselves as jewellers and opticians, able to deal in both spectacles and eyeglasses, as well as a range of fancy, decorative goods. In 1846, for example, Thomas Sale 'Jeweller & Optician' advertised in

The Bristol Mercury a stock of 'several hundred pairs of spectacles, eyeglasses, &c', alongside a variety of other items.[28] The account books of Robert Sadd, an 'Optician and Jeweller' from Cambridge, reveal the close connection between the jewellery trade and spectacles between 1837 and 1851. In five sample years, eyeglasses – a form of frame without side-arms – as well as spectacles and spectacle cases, were purchased consistently throughout the year, though not the primary form of income for the business.[29] Other traders styled as jewellers or engravers, but not opticians, also partook in spectacle retail. Several advertisements indicate that 'jeweller and watchmakers' or 'jewellers, watchmakers and importers of fancy goods' stocked and sold a range of spectacles and eyeglasses as part of their trade.

The separation between the fancier gold and silver frames and the far greater number of large, utilitarian, steel frames in the Science Museum's collections offer a visual demarcation of the early Victorian spectacle market. The introduction of cast steel was transformative and altered the sites of spectacle dispensing. Cast steel enabled a device that was increasingly functional, had an aesthetic vogue and could be manufactured more cheaply.[30] In both London and provincial towns, traders in cheaper metals and sundry goods stocked various types of thick, cast-steel spectacles. An early 'Tradesman's Account and Memoranda Book' of Samuel Brookes, a cutler and stationer from Wellington in Somerset, for example, demonstrates that customers purchased spectacles as well as a variety of sundries at different points throughout the year 1800.[31] At this lower end of the spectacle market, the trader did not need to advertise as a specialist in metalwork or optics, they just needed to emphasise the existence of a large range of stock. The early nineteenth-century bazaar, for example, was a fruitful space for the supply and distribution of spectacles, where they exhibited alongside an array of domestic and utilitarian miscellaneous goods.[32]

The appearance of spectacles in London and provincial advertisements, alongside plated goods, scientific instruments and cheaper, everyday products sold as part of the miscellaneous stock of a sundry goods trader, illustrates their growing commercial appeal by the early Victorian period. In the absence of any significant additional evidence about the businesses that supplied spectacles, where they are available trade cards and trade catalogues are an invaluable source for ascertaining the position of visual aids in the wider trade of the scientific instrument maker.[33] Historically, the importance of spectacles to the opticians' trade is best illustrated in their iconography, illustrated in Figure 1.1 below. Before the advent of street numbers, opticians were frequently found at the 'sign of the royal spectacles', or the 'two pairs of golden spectacles'.[34] In the Science Museum's collection of illustrated trade cards, spectacles and eyeglasses appear as a visible symbol of the optician, both for those who specialised in optical manufacture, and those who produced mathematical and philosophical goods.[35] Although on the trade card itself spectacles might occupy a marginal position – an adjunct obtained from a wholesaler in comparison to the various scientific apparatus that the trader manufactured or sold – spectacles were often the 'everyday' or quotidian item, part of the instrument maker's more domestic market, and held the potential to be the 'bread and butter' of their income.[36] The broad potential usership of spectacles might partly explain why they appear among the stock of the early Victorian instrument maker, because they would have enjoyed more universal appeal than specialised scientific instruments. Indeed, the difference in cost between the two was substantial; spectacles were much more affordable.[37]

Spectacles had a ubiquitous quality that was unusually adaptable and enabled them to fit within the remit and stock of a growing number of retailers. Surviving stock lists suggest that traders did stock the spectacles they advertised, and often in large quantities. The stock list accumulated at the death of William Strange, a watch and clock maker, in the early 1850s showed that spectacles and eyeglasses formed a part of an assortment of other goods, including combs and brushes, bronzed goods, gold rings, watches, clocks, clock movements, new plated silver and walking sticks.[38] Auction advertisements from newspapers and periodicals similarly demonstrate an equally varied range of stock at the time of an optician's death, retirement or bankruptcy. In the *Leeds Mercury* in February 1850 the 'Stock-In Trade of an Optician' included:

Figure 1.1 Trade card of A. Mackenzie, an optical, mathematical and philosophical instrument maker. Dated between 1816 and 1822, it reveals the historical prominence of spectacles in a scientific instrument maker's premises and window display.

Source: Science Museum Art collection, 1951-685/50. With permission from SSPL/Science Museum.

... upwards of 300 pairs of spectacles, several gross of cases, pocket glasses, telescopes, and multiplying glasses; dials, magnets, galvanic apparatus and plates; magic lanterns with glasses, thirteen barometers (unfinished), thermometer frames, a large quantity of lenses and glasses with a variety of other articles partly finished.[39]

The trade in spectacles alongside a sophisticated scientific instrument, or a finely crafted decorative piece of metalwork and a miscellaneous assortment of chess pieces and inkstands indicates that, in the plethora of retailing sites, the quality and design of frames and lenses varied substantially. In the eyes of early Victorian contemporaries, spectacle hawkers were positioned at the lowest end of this spectrum. In historical work the street has long been considered an important location for the sale of spectacles.[40] Street selling remained a popular early Victorian practice and for a range of trades – including the scientific instrument trade – the method of hawking via itinerant pedlars persisted well into the middle of the nineteenth century.[41] While it is difficult to quantify 'beyond the shop' retailing practices, John Benson and Laura Ugolini have argued that this does not lessen their importance or mean that shops were the dominant form of supply.[42] The descriptions of street vendors of spectacles, although anecdotal, provide some of the most vivid accounts of traders associated with spectacle sale in this period. Henry Mayhew's 1851 documentation of street sellers and performers in his account of *London Labour and the London Poor* placed spectacles in the category of 'manufactured articles'. Within this category, spectacles were a quotidian item situated in the ambiguous class of 'miscellaneous articles of manufacture' alongside a variety of items such as 'cigars, pipes, snuffboxes, combs... sponges, wash-leather, paper-hangings... [and] pin-cushions'.[43] The accuracy of Mayhew's account does need to be treated with caution, but the sale of spectacles could help streets sellers earn a living:

> There are sometimes 100 men, the half of whom are Jews and Irishmen in equal proportion now selling spectacles and eyeglasses... Some of these traders are feeble from age, accident, continued sickness or constitution and represent that they must carry on a 'light trade', being incapable of hard work even if they could get it.[44]

Here, desperation qualified a person to retail spectacles. Its implications are two-fold, highlighting first that expertise was not necessarily required in the dispensing of spectacles, and that they could be obtained cheaply and sold on, and second that the demand for spectacles was enough to provide street sellers with an income. Mayhew proposed that spectacle sellers did not necessarily 'confine themselves' to the sale of spectacles and sold anything that they thought was profitable.[45] This suggests that, although the profitability of spectacle selling could fluctuate, it was lucrative at certain times.

Warnings against the street selling practices described in Mayhew's accounts indicate that contemporaries judged the type of service offered, and quality of product distributed, by street sellers to be inferior. Cautionary tales indicate the geographical spread and putative conduct of itinerant and more fixed pedlars both in and outside the capital. Warnings against street sellers appeared in retailers' advertisements in, for example, Leeds, York, Dublin and Preston.[46] While the criticism of other spectacle sellers could have been part of a trader's method to market their own products, it does seem that the overall quality of the frame or lens varied at this time. The optician stood at the forefront of these charges of poor quality and published their critiques in treatises and pamphlets as well as advertisements. In 1838, extracts from George Cox's *Spectacle Secrets* appeared in two periodicals to 'expose' the widespread nature of fraud and the sale of substandard goods.[47] A year later, Thomas Harris & Son, opticians in Bloomsbury Square, London, claimed in their optical treatise that 'thousands' had had their eyes ruined because of fraudulent street sellers.[48] While both statements could have been exaggerated, revealing or eliminating street sale was on their agenda and the practice was documented to be multifarious. Moreover, although rural areas have often been associated with the street peddling of spectacles, it is clear through the accounts of Mayhew, London opticians and London-based periodicals, that it persisted in both the metropolis and provincial towns, in the early Victorian period.

For Thomas Harris & Son, the appropriate alternative to the street seller was the optician of 'experience and respectability'. Yet criticism of the street sale of spectacles is not as simple as pointing to the differences between 'shop' and 'beyond the shop' quality of goods and dispensing methods. 'Opticians' themselves were a diverse category of trader. Travelling opticians highlight the inability to clearly demarcate between these 'shop' and 'beyond the shop' practice and they travelled in ways not dissimilar to those of the itinerant medical quack.[49] A circular from a travelling optician, for example, explicitly states that he was not a 'pretender', but his position as a retailer is ambiguous because he both was travelling and advertised a permanent address.[50] Like pedlars, the credibility or reputation of travelling opticians with a fixed address might have been questionable. Travelling opticians' advertisements frequently adopted techniques that were open to criticism or suspicion, such as using the phrase 'licensed hawker' or claiming a 'new' invention. Similarly, Benson and Ugolini have argued that, across retailing as a whole, there was not a straightforward hierarchy and shop premises were not necessarily superior to street sale.[51] Nevertheless, for the more reputable optician, shop premises were an important part of their advertising strategy. Travelling opticians in their advertisements sought to separate themselves from street sellers and from suspicion through emphasising a permanent address as well as any temporary business address that they currently used. John Solomon, an optician and spectacle maker from Bristol, for example, advertised prolifically and emphasised his 'London and Bristol Optical Establishment'.[52]

Having a permanent establishment was meant to project an image of expertise, respectability and a certain quality of product. As Irvine Loudon has argued, a permanent address offered a customer some reassurance that a trader's claims had some substance – because they could not distribute wares falsely and then escape the town – and that they were capable of developing reliable relationships with their business clientele.[53] The marketing of visual aids can be placed in the wider context of 'polite commerce' that was established in the eighteenth century. Helen Berry and Jon Stobart, for example, have demonstrated the role of advertisements in helping to forge connections between the social elite and tradesmen and the growing aspirations of the middling classes.[54] Moreover, Stobart has also proposed that the diverse range of goods found in advertisements from the eighteenth century can be seen to represent the growing commercial world, as well as the trader's attempt to present the reader with a 'cornucopia'.[55] Opticians drew upon traditional forms of patronage and adopted the language of 'polite commerce'. Their trade literature is littered with imagery, crests and elaborate backgrounds as tools for enhancing their reputation and respectability, an integral component of nineteenth-century retail more generally.[56] Highly visible crests acted as an assured symbol of quality, and contributed to the 'democratisation of luxury' by allowing individuals to obtain products that were associated with elite society.[57] Both opticians' advertisements and their trade cards reference the monarch, members of the royal family and high-status organisations. The use of a royal crest or 'high society' connections bolstered the authority of an optician as vendor of the goods they sold, including spectacles, and presumably helped them to market their products.

A select group of instrument makers, including Hudson and Harris & Son, departed from opticians' traditional methods during the early Victorian period. Instead, they sought to market their products by explaining to their readers that spectacles were the focus of their trade and not just an adjunct or 'impulse-buy' section of their stock. An explicit example can be seen in Figure 1.2, the trade card of a London 'practical optician' that has been dated between 1840 and 1844. S. Phillips included a large image of spectacles and a price list, but only a short statement about other instruments: 'N.B. Barometers and Thermometers made and repaired'.[58] Phillips argued that, because of his 'many years' experience in the practical part of WORKING LENSES, he can with confidence recommend the use of his IMPROVED SPECTACLES'. The claims on these trade cards are corroborated by advertisements in the British Library Newspapers archive. A similar emphasis occurs in an advertisement for practical and mechanical opticians in the *North Wales Chronicle* in the 1840s, which discussed 'A New Discovery for the Eyes', and focused on the properties of their lenses, and their knowledge of the 'imperfections of sight'.[59] Such advertisements might suggest an increasing specialisation in the manufacture of spectacles by both individual retailers and wholesalers, to establish themselves as unique within the spectacle market. Morrison-Low has suggested that by the mid-nineteenth century spectacle making itself had become a separate enterprise from the

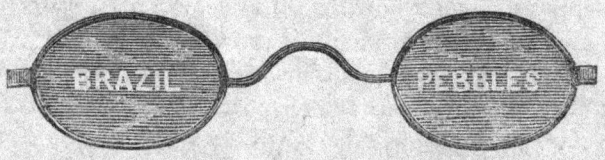

Figure 1.2 Trade card of S. Phillips, practical optician, with spectacle iconography and a focus on spectacles and vision expertise. Dated between 1840 and 1844.

Source: Science Museum Optics collection, 1921-323/137. With permission from SSPL/Science Museum.

scientific instrument trade. The evidence for this lies in trade directory records where a trader's profession evolved from, for example, an 'optician' in 1815, to an 'optician and spectacle maker' in 1830 and finally a 'spectacle maker' by 1850.[60]

Opticians' treatises produced by those specialising in spectacle sale became an important part of their claims to expertise in a variable retailing landscape that supported opticians of various classes and spectacle sale from several different types of trader. The treatise was a 'common mode' of marketing for new medical inventions and assistive devices because it offered the ability to publicise new technologies to potential clients and establish a sense of trust in their quality or utility.[61] Opticians built upon a longer tradition of optical and medical writings that had discussed spectacles and spectacle use since the eighteenth century. Medical personnel distributed information on spectacles as part of medical dictionaries, general medical works and family health guides.[62] Optical texts covered the same content and, in the absence of author information, it would be difficult to discern their trade or profession; both forms of text explained the basic principles of lenses for the treatment of two known vision defects defined as 'presbytae' (old sight) and 'myopia' (short sight).[63] By the early Victorian period opticians were producing a more expansive range of popular literature to provide information on two key areas: a continued explanation of vision defects, and new emphasis on the practicalities of spectacle wear. Optical texts informed their readers of the anatomy of the eye and principles of refraction, to aid understanding of how certain anomalies affected visual acuity, and how lenses could remedy this. Authors sought accessibility and ways to explain complex ideas visibly and clearly. Thomas Harris & Son's 1839 text and London optician Francis West's earlier text in 1827, for example, included a range of optical diagrams to illustrate the refractive properties of both the eye and glass lenses.[64] Authors also claimed a wide readership for their work. In 1835 an entry in the *Monthly Magazine* for London optician John Harrison Curtis's work reported that his text had sold '4,000 copies... in a short time'.[65] Others adopted a practice of thanking the public for buying their text in subsequent editions. In 1855, for example, London optician Charles Long in his preface 'expresses his gratification at the favourable reception experienced by the first issue'.[66] Long also included opinions of his first edition from the press, and while these two methods do not provide proof of a wide readership, the reviews from the popular press covered a wide geographical area.[67] Indeed, the similar spread of popular literature in America suggests that there was an Anglo-American appetite for advice on ocular health and how to preserve vision.[68] The range of reviews and extracts appearing in periodicals and newspapers in Britain indicates that opticians' ideas were being published in media capable of influencing public opinion and reflect the optician's growing public role in meeting popular demand to advise on the refractive state of the eye and use of spectacles.[69]

The language used in early Victorian opticians' treatises was anatomical. The rise of physiological optics blurred the lines between medical practitioners' and opticians' work and their public roles.[70] The titles of opticians' treatises demonstrate the overall function of the texts, to evidence their claims of a combined optics and anatomical expertise. Thomas Harris & Son's publication was titled *A Brief Treatise on the Eyes, Defects of Vision, and the Means of Remedying the Same by the Use of Improper Spectacles* and Francis West's *A Familiar Treatise on the Human Eye: Containing Practical Rules that will Enable all to Judge what Spectacles are Best Calculated to Preserve their Eyes to Extreme Old Age*. Within these, the nature and structure of the eye is explained.[71] In publishing this knowledge, and in their claims of expertise, these opticians sought to educate their popular readership and protect their eyes from damage incurred from spectacles bought at inappropriate locations. West, for example, emphasised the urgency and necessity of 'applying to regular opticians' as opposed to spectacle hawkers or other shops.[72] In 1827 the firm argued that the optician was superior because of their knowledge of the science of optics. However, by 1839, Harris & Son argued that both optics, *and the correct knowledge of the structure of the eye*, were necessary requisites for the ability to supply appropriate spectacles and remedy 'deficiency of vision'.[73] In the early Victorian period, opticians increasingly sought to differentiate themselves from the broader body of the spectacle retail trade by highlighting their understanding of scientific optics and anatomical and physiological knowledge of the eye. To return to John Thomas Hudson's treatise, discussed at the

Early Victorian understandings 57

outset of this chapter: Hudson was part of a body of opticians utilising the language of physiology and optics in their public position as authorities on the fitting of spectacles. In doing this, spectacles were presented as a sophisticated and complex device, valued for its sight-enhancing properties, and cautions were expressed because of its potential to damage the eye through misapplication.

The practice of spectacle dispensing and medical alternatives

The optician's claim of expertise came not necessarily from their ability to manufacture quality products, but from their ability to understand the human eye, recognise lenses and frames of quality and supply them appropriately. The optician was unable to differentiate their dispensing practice from those of other retailers. All retailers relied on trial and error; a customer would try on lenses of different gradations to find the one that suited them best. The optician claimed to guide the accuracy and credibility of this sale transaction. Broader concerns in retail, including false advertisement and persuasion, were of high importance in the sale of devices or therapeutic remedies that could, and would, be applied to the body. However, spectacles were distinct from other types of assistive device in terms of the timing and content of their associated advertisements and dispensing practices. Whereas other sellers of corrective body devices – for example steel trusses – had aligned themselves and advertised connections with medical institutions or personnel since the eighteenth century, opticians did not begin to consistently consider or claim any form of medical connection until the 1830s.[74] The sale of spectacles was distinct and responded to changes in ophthalmology that led to the cultivation of a specific form of expertise, one that was increasingly anatomical as well as scientific. As evident in opticians' texts, it was this combined skillset that differentiated the optician from other groups of traders and justified their claims of expertise. This expertise increasingly allied with, *and* established itself as distinct from, medical practice.

The early discipline of ophthalmology at the outset of the nineteenth century engaged with theories on the use of spectacles but had a negligible impact on spectacle retail and use. In 1793

Dr William Rowley, member of the Royal College of Physicians and physician to the St Marylebone Infirmary, produced a text on the eyes and eyelids which three years later formed part of his multivolume treatise on the *Rational Practice of Physic*. Rowley's treatise was the outcome of twenty years' work on the eye in response to 'defective' methods of treating its diseases and the 'pretensions of itinerant oculists, and the neglect of regular practitioners'.[75] He argued in his work that spectacles needed to be treated with caution but 'were necessary to rectify defects of vision' if these originated from 'peculiarity in the figure of the eye or advanced age'.[76] While Rowley's work highlights that spectacles had a place in the emerging discipline of ophthalmology, a much greater emphasis was placed on explaining and treating ophthalmic diseases. Such a focus characterises the trend in British ophthalmology in the early nineteenth century following the foundation of the London Eye Infirmary (later Moorfields Eye Hospital) and several specialist institutions in the 1810s. Ophthalmologists sought to establish a credible sub-discipline, one that was based on remedying eye disease through therapeutic, clinical principles, in order to differentiate their practice from the travelling 'oculist' or general practitioner.[77] In Britain, ophthalmologists were further tasked with overcoming medical practitioners' heavy mistrust of specialisation.[78] This is not to say that early ophthalmologists did not consider the use of spectacles. Spectacles appear in the textbooks that were written in the 1810s in an attempt to provide an authoritative body of knowledge for a growing number of students beginning to specialise in the rapidly changing discipline in Britain, Europe and America.[79] In 1815 for example, Georg Beer, a pioneering Austrian ophthalmologist, advised on the misuse of spectacles but did not 'entirely forbid the use of glasses'. One of Beer's primary reasonings for this was because, in his words, 'I know it will be attended to'.[80] From the outset ophthalmologists were aware of the crossovers between their practice and spectacle use. However, ophthalmologists generally cautioned against the adoption of spectacles and devoted their time to diagnosing, explaining and treating ophthalmic disease.

Opticians constructed their form of expertise in response to internal and external pressures derived from their awareness of medical practitioners' concerns about the role of quackery in the treatment of the eye and eye diseases. At the forefront of these pressures was the persuasive rhetoric and honesty of other retailers and the questionable quality of retailed goods. Opticians considered the conduct of the seller to be paramount because of the almost uniform method among retailers of dispensing through trial and error, where the emphasis lay on the customer choosing their own lens. The autonomy of the customer in the dispensing process placed greater emphasis on the need for the retailer to monitor both the quality of goods and the method of sale. Most work on how spectacles have been dispensed in the past has been written as part of the broader history of optometry. These studies tend to document the early tradition of trial and error, while looking at a few particularly novel ways in which certain individuals tried to create objective methods.[81] In her history of the British Optical Association – a collective organisation of vision-testing opticians established in the 1890s and discussed in Chapter Four – Margaret Mitchell concludes that 'before 1880 the hawker's tray and the empirical advice of the optician was the best service that could be offered'.[82] She suggests that there was little difference between the methods of those dispensing on the street and in shops. Moreover, the early nineteenth century is generally believed to have been a period of little change, because advancement relied on more sophisticated testing and diagnostic equipment that was invented from the mid-century onwards.[83] In many ways, these suppositions are accurate, and all retailers did utilise the trial-and-error method. However, the period was one of greater change than has previously been acknowledged and early inventions paved the way for the better documented later advancements. The rhetoric and discussion of vision testing among retailers highlights that the values and expectations of a retailer differed substantially even if differences in the methods of dispensing seemed negligible. Moreover, the exact nature of the 'trial and error' method of dispensing lenses has not been previously explored and this is likely because of the lack of clear evidence concerning the practices adopted by traders in this period. The amalgamation of a variety of fragmentary accounts in business archives, advertising, popular literature and, more rarely, advice manuals written by opticians and members of the medical profession is telling. Opticians sought to cultivate a method of sale that was different and, in doing so, established an area of public expertise.

Opticians condemned and criticised the trader whose conduct stood them the furthest apart from the opticians' practice: the street seller. A negative stereotype of spectacle hawkers' dispensing methods prevailed; they were claimed to coerce their customers and falsely claim new inventions that would improve a person's vision. As outlined in the previous section, in their advertisements opticians adopted a strategy of warning their potential customers against such 'pretenders'. In 1842, E. Solomons, an optician advertising in a Dublin newspaper, cautioned his customers against 'parties who assume his name, and travel from town to town, vending spectacles calculated to injure the sight to a frightful extent'.[84] Solomons's caution was not simply an advertising ploy; a fortnight later a former employee, having been dismissed from service, assumed the name of 'H.C. Solomon' and was advertising in local newspapers using the same testimonials as his previous employer.[85] Four years later, it appears, the same individual continued to advertise in a different locality, forcing E. Solomons to instigate criminal proceedings against him.[86] Popular stereotyping also presented hawkers negatively. In a provincial periodical, the persuasive methods of street sellers on one of the town's streets were described in aggressive terms, as 'thrusting their wares on the walking public'.[87] For the sale of spectacles, however, it was urged that damage could be caused by falsely utilising the rhetoric of the more reputable optician. In 1831 a letter written by Emma Botham Alderson – sister to the prolific nineteenth-century Quaker writer Mary Howitt – provides a striking description of an encounter with a spectacle hawker, remarking that:

> [h]e proved to be the most truly... worthiest old man I ever met or hope to meet... really philosophical[,] he talked of the laws of light & colours & the formation of the human eye most beautifully and yet so quaintly, that I could hardly refrain laughing at times & yet I listened in astonishment and pleasure.[88]

At first glance the letter indicates that customers harboured negative stereotypes or assumptions about street sellers. The seller is transformed in Emma's writing from her first impression of 'coarse features & mean appearance' into her later observations of a 'poor disguised wayfarer' with 'noble intellect'. The letter thus more subtly documents the methods street sellers used to attract their customers and the potential success of drawing upon both optical and ocular knowledge. It might be that the seller possessed this knowledge legitimately. Indeed, Benson and Ugolini have highlighted the difficulty of assessing the conduct of street sellers through the eyes of their critics.[89] In Henry Mayhew's accounts, for example, one seller did describe himself as an optician out of work.[90] However, Mayhew also recorded a conversation with a spectacle street seller who concluded that their practice, and incorporation of a language of expertise, perhaps driven by desperation or by keenness to make profit, was almost a matter of coercion:

> I think it's more in the way of persuasion... Why, I've persuaded people, when I was in the trade and doing well at it – for that always gives you good spirits – I've persuaded them in spite of their eyes that they wanted glasses. I knew a man who used to brag that he could talk people blind and they bought.[91]

The methods adopted by street sellers can be connected to the wider issue of persuasion in medical quackery, and the dispensing of medical goods using advertising and sales pitches. Indeed, the ability of street sellers to 'talk people blind' is not dissimilar to that of medical quacks of this period who could 'out-argue illness'.[92] In doing so, they had the potential to raise the standing of the street seller and lower the credibility of the optician, as well as potentially threaten the 'sanctity of one's sight'.[93]

Methods adopted by a broader range of spectacle traders can also be connected to another method of medical quackery, namely the adoption of press advertising and its potential to promote false or exaggerated statements. In 1839 an advert for 'West's Improved Stanhope Lens' claimed that 'several unprincipled traders have been styling themselves OPTICIANS', therefore suggesting that the use of titled professions in contemporary advertising was unregulated.[94] This has been shown to be the case in studies of medical advertising by Hannah Barker and Lisa Forman Cody, which have demonstrated that 'quacks' adopted persuasive techniques to sell their wares under limited regulation.[95] West's use of capitalisation also mimicked the adoption of large and bold typefaces by traders to grab readers' attention and has been increasingly linked to 'puffing'.[96] The early Victorians, as a result, began to question the moral responsibility of the public press in representing those dispensing spectacles or claiming expertise in diseases or conditions of the eye. In 1839 a medical lecture on diseases of the eye delivered at the Birmingham Royal School of Medicine and Surgery by Richard Middlemore, surgeon to the Birmingham Eye Infirmary, argued that the press should cease to 'pollute their pages and prostitute their pens' with the advertisements of empirics.[97] Readers of the popular press were equally aware of the poor dispensing practices. In 1843, for example, 'one who is not to be duped by quackery' wrote to *The York Herald* to explain that 'the system of puffing extortion so generally and successfully' adopted by several spectacle dispensers was a 'daily practice'.[98]

The type of criticism retailers faced, and the presence of spectacles and eyeglasses amongst the stock of a wide range of retailers, suggests that the dispenser did not necessarily require knowledge of the eye, optics or lens grinding. It also proposes that opticians provided a different type of service, one that medical practitioners did not consider to be polluting the pages of newspapers. At first glance, the legitimacy of opticians' claims appears questionable and raises the potential need for them to be placed under considerable scrutiny. Opticians, for example, prolifically advertised in this period. But written accounts and advice through postal correspondence on how vision could be tested reveal the finer details of dispensing. Indeed, while the detail of a remote sale transaction could appear minimal, opticians justified the credibility of their own practice on their ability to guide this process and ensure it was accurate. In 1839, Thomas Harris & Son, for example, stated in their treatise that they required the following information: whether the person had previously worn spectacles, and for how long; the distance they could see best; and the intended purpose.[99] Other opticians offered a similar service in their advertisements, highlighting that customers could describe their current visual acuity and be forwarded trial lenses to ascertain the lens that they felt fitted best.[100] Postal correspondence instructions show that the commonly asked questions were almost universal: they required

Early Victorian understandings 63

customers to provide the lenses previously used, the length of previous wear, the distance at which customers could read certain print and whether customers intended to use spectacles for near or distance work.

Brownlee in his work on antebellum America draws upon John Thomas Hudson – the London optician introduced at the outset of this chapter – and his sight testing methodology to demonstrate that spectacles could only be dispensed accurately in collaboration with the prospective user.[101] Hudson also distributed self-testing kits. A box labelled 'The Complete Sight Suiter: Or Book of Lenses by John Thomas Hudson', and dated to 1854, can be found amongst the test spectacles in the Science Museum's Ophthalmology collection.[102] The box contains ten cardboard-framed lenses, each marked 'J.T. Hudson, Optician, London' on one side, that have different degrees of optical strength that are visibly, and obviously, discernible. Printed on the other side of each lens is 'prices 10/6, 12/3, 1/6: The Complete Sight Suiter, sold by all booksellers', which suggests that this kit was intended to be used as a self-testing device by the prospective customer, and was not necessarily intended for spectacle dispensers. For Hudson, placing the customer in complete independence of the retailer was not haphazard providing that it occurred in the right context and under appropriate guidance. Indeed, he concluded that 'no optician however great his natural genius, or acquired skill can be half so good a judge of all these matters as the intelligent wearer himself'.[103] Moreover, he claimed that customers could 'judge accurately' when applying for frames from a catalogue and 'at a distance' from the shop.[104] However, while the customer held greater autonomy in choosing an appropriate lens, Hudson argued that the overall fit of the frame was important and half an hour should be put aside to ensure precision and accuracy.[105] The existence of Hudson's sight tester, and his discussion of the vision testing process, challenges the conception of self-testing and trial and error as a solely haphazard and random method of sale. This testing kit may seem arbitrary to a modern user, but Hudson perceived it to be the best method of dispensing because he judged that the user was in the best position to decide which lenses suited their eyes. While all retailers adopted methods of trial and error, opticians purported to be able to recognise the danger of a lens being too strong or weak and to fully understand the damaging effect that it could have. The risks associated with the use of incorrect lenses heightened the importance of accuracy, and accelerated opticians' and certain medical practitioners' early attempts to improve the vision testing process. The development of instruments to improve the objectivity of vision testing was not new. In 1801, Thomas Young, a medically trained polymath, adapted the optometer – an instrument designed to measure the eye's refractive error and determine the strength of optical lenses required – as explained in his Royal Society lecture on the mechanism of the eye.[106] Young's attempts can be seen as part of broader moves within medical practice to incorporate a variety of diagnostic instruments – beginning with the stethoscope in 1819 – to measure the body in objective ways and reduce the reliance on the patient.[107] But such instruments were not initially introduced into ophthalmologists' practice and were slow to appear in retailers' advertisements, starting in the 1830s. In 1835, for example, B. Salom advertised his newly invented instrument called the 'Optician's Guide' in the *Liverpool Mercury*.[108] Although Salom argued that 'the dealer in Optics' rarely 'studied the nature and physiology of the eye', he emphasised his own training under a variety of 'talented and eminent gentlemen'. He argued that his instrument – which followed a similar principle to the optometer – was able to ascertain the focus of an individual's sight and remove 'all doubt... so as to preclude the least shade of error'. Opticians and instrument makers also sought to disseminate other types of measuring equipment to ensure an accurate fit of both the frame and lenses. In 1837 at the seventh meeting of the British Association for the Advancement of Science in Liverpool, for example, the importance of a correctly fitting frame was suggested in a paper 'On measuring the eyes for suiting them with spectacles'.[109] Here, it was argued that the width of the eyes differed between people and therefore should be accurately measured. It also recommended that an examination of the eye, by means of a card of 'twenty-four, equidistant, radial lines' to ascertain the longest and shortest focus of the eye, should be conducted before attempting to select spectacles.[110]

The push towards more objective methods highlights the alignment between opticians' responses to both internal demands to claim their legitimacy and external changes and expectations to improve vision testing. From the 1830s opticians sought to establish a firm connection with appropriate medical personnel and medical institutions. Early references to medicine highlight how the retailer was 'favoured', 'recommended' or 'patronised' by medical practitioners.[111] The authenticity of these statements could be questioned. Opticians therefore sought to combat this by providing increasingly more specific medical testimonials in their advertisements and keeping abreast of recent developments in the growing discipline of ophthalmology. Knowledge of the expanding number of ophthalmic institutions, for example, had a direct effect on the language used by opticians to attract customers and the type of testimonies they sought.[112] Traders set these testimonies next to a set of advertised skills, which placed a greater emphasis on anatomical and physiological knowledge at the expense of an earlier emphasis on length of practice and practical experience. It was not new to claim an ability to 'preserve' the sight or care for the eyes. Eighteenth- and early nineteenth-century trade cards and advertisements, for example, include claims that retailers could 'suit the sight' to prevent damage.[113] However, in these examples, advertisers did not claim any physiological understanding of the body. In contrast, by the 1830s, retailers both advertised and produced treatises on their anatomical and physiological expertise, which went beyond the quality or construction of the frame and lens of the spectacles sold. As Liliane Hilaire-Pérez and Christelle Rabier have acknowledged in their study of steel trusses, an established alliance between the medical faculty and a particular trader could have several benefits, including support for the authenticity of their claims, improvement of the devices sold and the fostering of working relationships with medical personnel.[114] Opticians, like Salom advertising in the *Liverpool Mercury*, sought these benefits by advertising their new innovations and their study of the anatomy and physiology of the eye and vision.[115]

Opticians positioned themselves as traders with a uniquely combined skillset at a time when ophthalmologists began to collectively establish and consider their own vision testing methods. In 1847, Alfred Smee, Fellow of the Royal College of Surgeons, in lectures delivered at the Central London Ophthalmic Hospital, discussed two measurements of the face and vision. First, his 'visometer' measured the face and visual axes so that the 'centres of vision may be learnt to one-hundredth of an inch without error'. Second, his 'optometer' resembled Salom's 'Optician's Guide', consisting of a graduated scale to obtain the point at which a person could see distinctly.[116] Smee, in further adapting the optometer, argued that the optical properties or defects of the eye could be ascertained with it. Although by the late 1840s practitioners had begun to incorporate diagnostic technologies into medical practice, at this point Smee did not propose that medical practitioners should use the optometer;[117] instead, he argued that it should 'invariably' be used by the optician before any spectacles were sold or chosen by the 'applicant'.[118] Discussion of Smee's optometer highlights the interchange between medical and scientific ideas that continued to delineate the roles of the optician and the emerging ophthalmologist and separate them along clear lines. Medical practitioners and ophthalmologists did not immediately adopt into their own practice the new technologies that were being developed and the optician remained the favoured option for determining what spectacles a person might require. In 1854 William Mackenzie, a surgeon and lecturer based in Glasgow and one of the founding fathers of British ophthalmology, expressed a less supportive view of the optometer's usefulness, arguing that to ascertain appropriate lenses for short sight 'the surest plan is to try a series of them at an optician's shop' and that the optometer should only be considered for those residing at a distance.[119]

Although anatomical knowledge was an important feature of the opticians' skillset by the 1850s, spectacle dispensing was divorced from medical practice and continued to be treated with caution. Smee, for example, argued that the wrong spectacles caused 'very great mischief' when incorrectly suited and that the potential for injury was often underestimated.[120] Moreover, while Mackenzie could refer a patient to the optician for spectacles for short sight, lenses were not the only approach considered; ophthalmologists might explore several treatment options in a bid to ascertain the most effective treatment method. Mackenzie described a case where a short-sighted person, eventually treated with spectacles, was first subjected to leeching, purgatives and blisters in an attempt to remedy their vision.[121] He did criticise this treatment and promoted the use of spectacles to remedy myopia and presbyopia (old sight). However, he also explored the value of exercising the eye when a person had myopia and suggested that concave lenses could 'aggravate' vision.[122] Additionally, Mackenzie advocated a variety of possible therapeutic treatments for a condition of the eyes known as 'asthenopia', which would later be remedied using optical lenses. 'Asthenopia' caused a person discomfort and a range of physical symptoms when they overused their eyes. Mackenzie suggested tonics, diet, sedatives, stimulants, spirituous and aromatic vapours, and even cauterisation of the urethra, in order to help alleviate symptoms and seek a cure. In instances where these were unsuccessful, he advised emigration to Australia, because such a change would enable the patient to undertake 'pastoral pursuits' and reduce the strain on their vision.[123]

Ophthalmologists' desire to adopt therapeutic methods could be based on the need to justify their role and develop remedies that aligned with their medical training. Mackenzie, for example, emphasised that an ophthalmologist's training did not include the use of lenses when he stated that a patient would need to attend an optician's shop. He was not dismissive of spectacles as a whole; he primarily based his reasoning on clinical experience and the use of therapeutic methods to treat a range of eye diseases and conditions that were presented in his vast treatise. As a result, the inclination to avoid the use of lenses in certain cases can also be explained by the constraints on medical knowledge and ophthalmological training at this time. 'Asthenopia', discussed by Mackenzie, is a good example of this, because effective treatment with lenses was dependent on being able to examine the eye in greater depth to ascertain the cause of the problem. Without the technology to achieve this, ophthalmologists adopted therapeutic methods, which were observed to have helped. In contrast, by 1850 the optician was increasingly positioned as a reputable trader who could observe the principles of optics, while being aware of medical knowledge of the eye, to fit suitable spectacles when therapeutic remedies were ineffective. A level of medical understanding had become an important feature of their business, as well as optical skill. In 1853 William Ackland, a London-based optician, philosophical and photographic instrument maker, advertised his 'medical knowledge as a Licentiate of the Apothecaries' Company, his theoretical knowledge as a mathematician and his practice as a working optician, aided by Smee's optometer'.[124] Opticians adopted the technologies developed by ophthalmologists to carve out a more distinctive dispensing strategy. In 1855 the London optician Charles A. Long praised Smee's 'optometer' while alluding to the changes that were taking place, and about to accelerate in the coming decades, that would overturn the increasingly established, but delicate and far from fixed, professional roles. The trial-and-error method of dispensing visual aids was the 'old plan' of a bygone era:

> The selection of spectacles and eye-glasses requires the greatest care and attention, and should not be performed hurriedly... The old plan of trying on a number of glasses is most injurious as the eye becoming fatigued refuses to perform its functions properly... We are indebted to Alfred Smee... the use of the Optometer, an instrument invented by that gentleman... enables the optician to determine at a glance, and without fatigue to the eyes of the applicant, the exact amount of correction that requires to be given to the sight.[125]

Conclusion

Spectacles to the early Victorian were a multifarious and miscellaneous device, which could be bought in a variety of places. Depending on the user or the intended function of the frame and lens, the early Victorian could visit an optician's shop to obtain a bespoke device, a jeweller to obtain a decorative and expensive metal frame, a bazaar to pick up a pair of spectacles alongside an assortment of sundries and/or they could purchase spectacles from a seller on the street. Being available in a variety of locations highlights the increasing ubiquity of spectacles. It demonstrates on one hand that they were an increasingly valued device, able to meet the visual needs of a range of potential customers. On the other hand, it also suggests that those who used them did not consider them to be particularly complex, or sophisticated. Two letters sent to a prominent and well-established London optician's firm, C.W. Dixey, provide an insight into the views held by some of their more genteel clientele. One from Lord Stamford in February 1843 indicated that a person's age could determine the lenses selected, and that no examination was required:

> Lord Stamford encloses Messrs Dixey his Evening Spectacles which do not now magnify sufficiently to read small print in the Morning. He therefore wishes to send him a Pair of Morning spectacles to magnify more, and a Pair of Evening spectacles great magnifiers in different coloured cases. Lord S is now in the Seventy Eight year of his age.[126]

Likewise, another letter dated 18 May 1852 suggested that a pair of spectacles could be chosen without examination, and based on the distance at which the wearer held his face from his plate at mealtimes:

> Lord Buckinghamshire will thank Mr Dixie to send him down to Sidmouth a pair of spectacles such as he thinks will suit a youth of 16 years of age who has never worn any but is so short sighted as to be obliged to hold his face close to his plate when he takes his meals. They are wanted for one of Lord B's Sons.[127]

Both customers believed spectacles could be dispensed appropriately from rudimentary information.

Reputable opticians, despite their claims of expertise, had to compete with traditional, popular beliefs and practices. The persistence of more haphazard or arbitrary beliefs amongst customers further exposes the nature of spectacle dispensing that has been outlined in this chapter. Up until this point spectacle dispensing remained, primarily, a standard commercial transaction, not dissimilar to the purchase of an ever-increasing number of goods being offered to the early Victorian consumer. Spectacle dispensing methods were governed by the broad early Victorian retail practices of marketing and selling, a landscape in which opticians were attempting to position themselves at the top. Opticians increasingly framed spectacles as a more complex device, one that demanded a specific skillset to be dispensed accurately and, if neglected, had the potential to damage a person's eyes. The values and expectations of spectacle dispensing shifted, even if popular perceptions or the methods of trial and error did not. This general trend runs parallel to broader developments in ophthalmology; opticians utilised specialised institutions and increased medical interest in the eye to bolster their claims. However, as highlighted by David M. Turner and Alun Withey in their study of eighteenth-century corrective body technologies, a trader advertising medical connections did not thereby mean that their device was conceived as a medical object.[128] Similarly, the practice of spectacle dispensing was far from medicalised in this period; practitioners often adopted other methods and spectacle dispensing helped to shape an optician's practice and professional role as distinct from medicine. Turner and Withey also argue that corrective body devices sought to provide aesthetic improvement as well as restore a person's functional capabilities. Did Victorian spectacles enhance a person's vision? Could spectacles be an object of display or improve a person's appearance? Did spectacles have a non-functional purpose? As we will see in the following chapters, the Victorians altered the functionality and visibility of spectacles as a device and shaped how these questions would be answered both at that time and in the present day. Spectacles became increasingly more functional, their retailing market became ever more competitive and their use ever more visible. The roots of this can be found in the early Victorian spectacle market and the actions of retailers who sought to portray and dispense spectacles as a lucrative and growing commercial product outside the remit of medical expertise.

Notes

1 J.T. Hudson, *Useful Remarks upon Spectacles, Lenses, and Opera-Glasses; with Hints to Spectacle Wearers and others; being an epitome of practical and useful knowledge upon this popular and important subject* (London: Joseph Thomas, 1840), p. 5.
2 Liliane Hilaire-Pérez and Christelle Rabier, 'Self-machinery? Steel trusses and the management of ruptures in eighteenth-century Europe', *Technology and Culture*, 54:3 (2013), 460–502; David M. Turner and Alun Withey, 'Technologies of the body: polite consumption and the correction of deformity in eighteenth-century England', *History: The Journal of the Historical Association*, 99:338 (2014), 775–796.

3 Turner and Withey, 'Technologies of the body', various.
 4 Alun Withey, *Technology, Self-Fashioning and Politeness in Eighteenth-century Britain: Refined Bodies* (Basingstoke: Palgrave Macmillan, 2016), p. 102.
 5 Richard Champness, *A Short History of the Worshipful Company of Spectacle Makers up to the Beginning of the Twentieth Century* (London: Apothecaries Hall, 1952), pp. 8–9.
 6 C.W. Dixey & Son, *A Short History, 1777–1977* (Newport: Mullock & Sons, 1977), p. 29 quotes several letters between Dixey and his customers. Sadly, the original letters were destroyed during the Second World War.
 7 Hilaire-Pérez and Rabier, 'Self-machinery?', p. 475.
 8 Turner and Withey, 'Technologies of the body', pp. 781–782; Hilaire-Pérez and Rabier, 'Self-machinery?', p. 462.
 9 See, for example, *Leeds Mercury* (1 August 1829); *The Bristol Mercury* (22 October 1853).
10 These findings have been compiled from the trade directory findings of project SIMON that are detailed in Gloria Clifton's *Directory of British Scientific Instrument Makers 1550–1851* (London: Zwemmer, 1995).
11 Of the objects in the collections studied, 76% display a London address in the 'signature'.
12 A.D. Morrison-Low, *Making Scientific Instruments in the Industrial Revolution* (Aldershot: Ashgate, 2007), p. 2; Michael Andressen, *Spectacles: From Utility Article to Cult Object* (Stuttgart: Arnoldsche, 1998), p. 285.
13 Morrison-Low, *Making Scientific Instruments*, p. 133; Clifton, *Directory of British Scientific Instrument Makers*, pp. xii and xiv.
14 Morrison-Low, *Making Scientific Instruments*, p. 133.
15 Daniel M. Albert, 'Ocular refraction and the development of spectacles', in Daniel M. Albert and Diane D. Edwards (eds), *The History of Ophthalmology* (Oxford: Blackwell Science, 1996), p. 115; C.W. Elson, *Origin and Development of Spectacles* (Worthing: Worthing Archaeological Society, 1935), p. 13; Richard Corson, *Fashions in Eyeglasses*, 3rd edn (London: Peter Owen, 2011), pp. 49, 66.
16 Ian Mitchell, 'Retailing innovation and urban markets, *c.*1800–1850', in Laura Ugolini and John Benson (eds), *Retailing Beyond the Shop, c.1400–1900* (Bradford: Emerald Group Publishing Ltd, 2010), p. 288.
17 See, for example, Albert, 'Ocular refraction', p. 115.
18 Ugolini and Benson (eds), *Retailing Beyond the Shop*.

19 See, for example, Kevin Williams, *Read all About It! A History of the British Newspaper* (London: Routledge, 2010), pp. xii; James Mussell, 'Elemental forms: the newspaper as popular genre in the nineteenth century', *Media History*, 20:1 (2014), 4–20; Tom O'Malley, 'The regulation of the press' in Martin Conboy and John Steel (eds), *Companion to British Media History* (London: Routledge, 2014), pp. 228–238.
20 Peter John Brownlee, *The Commerce of Vision: Optical Culture and Perception in Antebellum America* (Philadelphia: University of Pennsylvania Press, 2019), p. 42.
21 Jon Stobart, Andrew Hann and Victoria Morgan, *Spaces of Consumption: Leisure and Shopping in the English Town, c.1680–1830* (London: Routledge, 2007), pp. 35–37.
22 Morrison-Low, *Making Scientific Instruments*, p. 15.
23 For an example of this see Science Museum Art collection, 1948–397/14, 'Trade card: A. Davis', dated 1815–1899.
24 *The London Dispatch and People's Political and Social Reformer* (27 January 1837); this is repeated in later advertisements in the same newspaper: 8 October 1837, 22 October 1837, 19 November 1837, 26 November 1837, 24 December 1837, 25 March 1838, 29 July 1838 and 9 December 1838.
25 *North Wales Chronicle* (15 August 1837, 22 August 1837); *The Bradford Observer* (4 January 1838, 11 January 1838); *Derby Mercury* (15 January 1840, 22 January 1840, 29 January 1840, 5 February 1840); *North Wales Chronicle* (23 June 1840); *Preston Chronicle* (10 October 1840); *The Blackburn Standard* (11 November 1840, 18 November 1840).
26 *The Essex Standard and General Advertiser for the Eastern Counties*, weekly between 24 May 1850 and 14 June 1850 and *The Bury and Norwich Post and Suffolk Herald*, weekly between 29 May 1850 and 19 June 1850.
27 Science Museum Ophthalmology collection, object number A680673.
28 *The Bristol Mercury* (21 March 1846).
29 Cambridge University Library, GBR/0012/Ms Add.5781–5783: Account books of Robert Sadd, optician and jeweller, for the years 1837, 1839, 1847, 1849 and 1851.
30 Turner and Withey, 'Technologies of the body', pp. 781–782; Hilaire-Pérez and Rabier, 'Self-machinery?', p. 462.
31 Somerset Heritage Centre, DD/X/RCH/1: Tradesman's account and memoranda book of Samuel Brookes, cutler and stationer, 1797. Whilst the instances were not numerous in this account book, neither was the trade in other items large, across the year as a whole.

32 *The Bristol Mercury* (12 September 1835).
33 Michael A. Crawforth, 'Evidence from trade cards for the scientific instrument industry', *Annals of Science*, 42:5 (1985), 453–544; Morrison-Low, *Making Scientific Instruments*, pp. 11, 226.
34 For examples of this on eighteenth-century trade cards, see Science Museum Art collection, 1951-685/49, dated 1776–1784; 1934-122/21, undated; 1934-97/2; 1934-106, dated 1706–1756; 1934-110, dated 1745–1772.
35 See, for example, Science Museum Art collection, 1934-121/131, dated 1812 and 1951-685/18, undated.
36 Morrison-Low, *Making Scientific Instruments*, p. 108.
37 Please see Chapter Three for an assessment of the cost of spectacles.
38 The National Archives, J/90/834: Catalogue of the 'late' William Strange, watch and clock-maker.
39 *Leeds Mercury* (20 July 1850); see a similar example in *The Athenaeum* (19 September 1853), p. 355.
40 See, for example, Corson, *Fashions in Eyeglasses*, p. 52; Elson, *Origin and Development of Spectacles*, p. 13; Carl Barck, *The History of Spectacles, Originally Delivered as a Lecture Before the Academy of Science* (Reprinted from the Open Court for April, 1907), p. 15.
41 Benson and Ugolini, 'Beyond the shop: problems and possibilities', in Benson and Ugolini (eds), *Retailing Beyond the Shop*, pp. 257–259; David Alexander, *Retailing in England during the Industrial Revolution* (London: The Athlone Press, 1970), p. 65; Morrison-Low, *Making Scientific Instruments*, pp. 205–206.
42 Benson and Ugolini, 'Beyond the shop: problems and possibilities', pp. 258–259; Alexander, *Retailing in England* also discusses the potential problems of quantifying street selling, pp. 61–65.
43 Henry Mayhew, *London Labour and the London Poor, A Cyclopaedia of the Condition and Earnings or Those that Will Work, Those that Cannot Work, and Those that Will Not Work*, vol. 1 (London: Griffin, Bohn and Company, 1851), p. 324; see also p. 3.
44 Ibid., pp. 444–445.
45 Ibid., p. 445.
46 See, for example, *Leeds Mercury* (1 August 1829); *Freeman's Journal and Daily Commercial Advertiser* (19 November 1842, 25 November 1842, 3 December 1842); *The York Herald and General Advertiser* (21 October 1843); *The Preston Guardian* (3 August 1844, 5 October 1844, 16 November 1844, 18 January 1845, 1 March 1845, 22 March 1845, 12 April 1845).

47 'The library table', *The Athenaeum* (1 December 1838), p. 855; 'Literary register', *Tait's Edinburgh Magazine* (December 1838), pp. 803–804.
48 Thos Harris & Son, *A Brief Treatise on the Eyes, Defects of Vision, and the Means of Remedying the Same by the Use of Improper Spectacles, Also Rules for judging when Spectacles are necessary, and Directions for selecting them* (London: Onwhyn, 1839), p. 32. See also Georg Beer, *The Art of Preserving the Sight Unimpaired to an Extreme Old Age; and of Re-establishing and Strengthening it When it is Become Weak* (London: Henry Colborn, 1815), pp. xiii–xiv; W. Kitchiner, *The Economy of the Eyes: Precepts for the Improvement and Preservation of Sight* (London: Hurst Robinson & Co., 1824), p. 7; and criticism can also be found in a popular text on spectacles by Andrew Ross, *On the Use and Abuse of Spectacles* (London: R. Kinder, 1840), p. 1.
49 Irvine Loudon, 'The Vile Race of Quacks with which this Country is Infested', in W.F. Bynum and Roy Porter (eds), *Medical Fringe & Medical Orthodoxy 1750–1850* (London: Croom Helm, 1987), p. 115.
50 For another example of this see Science Museum Art collection, 1948-397/14, dated 1815–1899.
51 Benson and Ugolini, 'Beyond the shop: problems and possibilities', p. 266.
52 Clifton, *Directory of British Scientific Instrument Makers*, p. 258; see, for example, advertisements in the *Southampton Herald* (28 January 1832, 12 August 1837).
53 Loudon, 'The Vile Race of Quacks', p. 118.
54 Helen Berry, 'Polite consumption: shopping in eighteenth-century England', *Transactions of the Royal Historical Society*, 12 (2002), 375–394 (pp. 383–4, 393–4).
55 Jon Stobart, 'Selling (through) politeness: advertising provincial shops in eighteenth-century England', *Cultural and Social History*, 5:3 (2008), 309–328.
56 See, for example, John Benson, 'Drink, death and bankruptcy: retailing and respectability in late Victorian and Edwardian England', *Midland History*, 32:1 (2007), 128–140.
57 Berry, 'Polite consumption', pp. 383–4, 393–4.
58 Science Museum Art collection, 1948-397/44, Trade card of S. Phillips, dated 1840–1844.
59 *North Wales Chronicle*, weekly between 7 May 1844 and 18 June 1844.
60 Morrison-Low, *Making Scientific Instruments*, p. 273.

61 Hilaire-Pérez and Rabier, 'Self-machinery?', p. 481; see also Claire L. Jones, discussing the evolution of the medical trade catalogue from the treatise as a form of information technology: 'Instruments of medical information: the rise of the medical trade catalog in Britain 1750–1914', *Technology and Culture*, 54:3 (2013), 563–599.

62 See, for example, Robert James, MD, *A Medicinal Dictionary, including physic, surgery, anatomy, chymistry and botany*, vol. III (London: T. Osborne, 1743–5), available at https://web.archive.org/web/20230122191938/https:/data.historicaltexts.jisc.ac.uk/view?pubId=ecco-0307200103&terms=Medicinal%20Dictionary,%20including%20physic,%20surgery,%20anatomy,%20chymistry%20and%20botany (accessed 22 January 2023), digital page 96; William Lewis, MB FRS, *Medical Essays and Observations, published by a society in Edinburgh, containing meteorology, mineral waters, material medica and pharmacy*, vol. I (London, 1746), p. 387; George Motherby, *A New Medical Dictionary; Or, General Repository of Physic* (London: J. Johnson, 1791), p. 528; F.M. Willich MD, *Lectures on Diet and Regimen: Being a systematic inquiry into the most rational means of preserving health and prolonging life* (London: A. Strahan, 1800), pp. 652–658.

63 William Emerson, *The Elements of Optics. In Four Books* (London: J. Nourse, 1768), pp. 156–167; Joseph Harris, *A Treatise of Optics: Containing Elements of the Science, In Two Books* (London, 1775), pp. 144–152; Erasmus Middleton, *The New Complete Dictionary of Arts and Sciences*, vol. II (London, 1778), digital page 408; Addison Smith, *Visus Illustratus; or, the sight rendered clear and indistinct* (London, 1783), throughout.

64 Harris & Son, *A Brief Treatise on the Eyes*, front page; Francis West, *A Familiar Treatise on the Human Eye: Containing Practical Rules that will Enable all to Judge what Spectacles are Best Calculated to Preserve their Eyes to Extreme Old Age*, 2nd edn (London: W. Ackrill, 1827), front page.

65 'Monthly review of literature – observations on the preservation of sight', *Monthly Magazine, or British Register* (September 1835), p. 315.

66 Charles A. Long, *Spectacles: When to Wear and How to Use Them: Addressed to Those Who Value Their Sight* (London: Bland and Long, 1855), preface.

67 Long included reviews from *The Banker's Circular*, *Ipswich Express*, *Hampshire Advertiser*, *Newcastle Guardian*, *Bedford Times*, *Sheffield and Rotherham Independent*, *Salisbury and Winchester*

Journal, the *Western Times*, the *Nottingham Review*, and the *Sussex Advertisement*.
68 Brownlee, *The Commerce of Vision*, pp. 6, 8, 10, 27.
69 See, for example, Francis West in 'Literature', *Operative* (27 January 1839).
70 Brownlee has identified the same in America: *The Commerce of Vision*, p. 22.
71 Harris and Son, *A Brief Treatise on the Eyes*, pp. 9–19; West, *A Familiar Treatise on the Human Eye*, pp. 11–13, 33–37.
72 West, *A Familiar Treatise on the Human Eye*, p. 26.
73 Harris & Son, *A Brief Treatise on the Eyes*, p. 7.
74 See Hilaire-Pérez and Rabier, 'Self-machinery?' and Turner and Withey, 'Technologies of the body'.
75 William Rowley, *A Treatise on One Hundred and Eighteen Principle Diseases of the Eyes and Eyelids* (London: J Wingrave, 1790), pp iii–iv; William Rowley, *Dr Rowley's Rational Practice of Physic, in Four Volumes*, vol. III (London, 1793), pp. i–ii.
76 Rowley, *A Treatise*, p. 353; Rowley, *Dr Rowley's Rational Practice of Physic*, p. 397.
77 Luke Davidson, '"Identities ascertained": British ophthalmology in the first half of the nineteenth century', *Social History of Medicine*, 9:3 (1996), 313–333.
78 George Weisz, 'The emergence of medical specialisation in the nineteenth century', *Bulletin for the History of Medicine*, 77 (2003), 536–575; George Weisz, *Divide and Conquer: A Comparative History of Medical Specialisation* (New York: Oxford University Press, 2006); George Rosen, *The Specialisation of Medicine with Particular Reference to Ophthalmology* (New York: Froben Press, 1944); Stephen Casper, *The Neurologists: A History of a Medical Specialty in Modern Britain, c.1789–2000* (Manchester: Manchester University Press, 2016), pp. 60–64. There was also some level of this in America: see Brownlee, *The Commerce of Vision*, p. 24.
79 Davidson, '"Identities ascertained"', p. 324.
80 Beer, *The Art of Preserving the Sight Unimpaired*, p. 206.
81 See, for example, Wolfgang H. Vogel and Andreas Berke, *Brief History of Vision and Ocular Medicine* (Amsterdam: Wayenborgh Publishers, 2009), pp. 202–239; Albert, 'Ocular refraction', pp. 108–119.
82 Margaret Mitchell, *History of the British Optical Association, 1895–1978* (London: British Optical Association, 1982), p. 19.

83 William Rosenthal, *Spectacles and Other Vision Aids: A History and Guide to Collecting* (San Francisco, CA: Norman, 1996), pp. 308, 311; Barck, *The History of Spectacles*, p. 20.
84 'Advertisements & notices', *Freeman's Journal and Daily Commercial Advertiser* (19 November 1842, 25 November 1842, 3 December 1842).
85 'Advertisements & notices', *Freeman's Journal and Daily Commercial Advertiser* (19 December 1842). He spelled his name 'Henry Cohen' with an 'e'; for later changes, see below.
86 'Advertisements & notices', *Freeman's Journal Daily Commercial Advertiser* (8 January 1846, 17 January 1846, 3 March 1846, 28 April 1846). What is interesting in this latter case, however, is the trader changing the spelling of his name in advertising, from 'Henry Cohan' to 'Henry Cowan', as an attempt to deceive.
87 'Pencilling of persons – No. III', *Bradshaw's Manchester Journal* (7 August 1841), pp. 235–237.
88 University of Nottingham Archives, Manuscripts & Special Collections, HT/7/1/30: Letter from Emma Botham Alderson to her sister Mary Howitt, 26 October 1831.
89 Benson and Ugolini, 'Beyond the shop: problems and possibilities', pp. 260–262.
90 Mayhew, *London Labour and the London Poor*, p. 444.
91 Ibid.
92 Roy Porter, *Health for Sale: Quackery in England, 1660–1850* (Manchester: Manchester University Press, 1989), pp. 94–96.
93 See Brownlee's discussion of America: *The Commerce of Vision*, pp. 61–63.
94 *The Athenaeum* (15 June 1839).
95 Lisa Forman Cody, '"No cure, no money", or the invisible hand of quackery: the language of commerce, credit and cash in eighteenth-century advertisements', *Studies in Eighteenth-Century Culture*, 28 (1999), 103–130; Hannah Barker, 'Medical advertising and trust in late Georgian England', *Urban History*, 36:3 (2009), 379–398.
96 Brownlee notes a similar phenomenon in America: *The Commerce of Vision*, pp. 89–92.
97 Richard Middlemore, *Introductory lecture on the Anatomy, Physiology, and Diseases of the Eye, Delivered at the Birmingham Royal School of Medicine and Surgery, 4 October 1839* (London: S. Longman, Orme, Brown, Green and Longmans, 1839), pp. 17–18.
98 'Spectacles', *The York Herald and General Advertiser* (21 October 1843).

99 Harris & Son, *A Brief Treatise on the Eye*, p. 26.
100 E. Solomons, *Freeman's Journal and Daily Commercial Advertiser*, regularly (almost weekly) between 14 January 1843 and 17 June 1843; Mr Abraham, *Hampshire Advertiser & Salisbury Guardian* (20 May 1837, 27 May 1837, 17 June 1837, 29 July 1837).
101 Brownlee, *The Commerce of Vision*, p. 58.
102 Science Museum Ophthalmology collection, object number A600351.
103 Ibid.; Hudson, *Useful Remarks upon Spectacles*, p. 28.
104 Hudson, *Useful Remarks upon Spectacles*, pp. 23–24, 28, 32.
105 J.T. Hudson, *Spectalaenia; or the Sight Restored, Assisted and Preserved by the Use of Spectacles* (London: Simpkin and Marshall, 1833), p. 29.
106 David A. Atchison and W. Neil Charman, 'Thomas Young's contribution to visual optics: The Bakerian Lecture "On the mechanism of the eye"', *Journal of Vision*, 10:12 (2010), 1–16; Dorothy M. Turner, 'Thomas Young and the eye on vision', in E.A. Underwood (ed.), *Science, Medicine and History: Essays on the Evolution of Scientific Thought and Medical Practice Written in Honour of Charles Singer* (Oxford: Oxford University Press, 1954), pp. 243–255.
107 Stanley Joel Reiser, *Medicine and the Reign of Technology* (Cambridge: Cambridge University Press, 1978).
108 *Liverpool Mercury* (17 July 1835).
109 'The British Association', *The Literary Gazette: A Weekly Journal of Literature, Science, and the Fine Arts* (7 October 1837), pp. 643–644.
110 Ibid.
111 See, for example, *Preston Chronicle* (10 October 1840); *The Blackburn Standard* (11 November 1840, 18 November 1840); *Freeman's Journal and Daily Commercial Advertiser* (19 November 1842); and the advertisements of Chamberlain in the *Examiner* between 15 April 1838 and 13 June 1840 (the latter: p. 487); *The London Dispatch and People's Political and Social Reformer* (27 January 1839); *Figaro in London* (11 February 1839), p. 47; *The Athenaeum*, between 30 November 1839 and 12 December 1840 (the latter: p. 999); and *The Musical World*, between 14 May 1840, p. 312 and 23 July 1840, p. 63.
112 See, for example, *Liverpool Mercury* (17 July 1835).
113 See, for example, Science Museum Art collection, eighteenth-century examples: 1934-97/2; 1934-121/43; 1951-687/19; 1951-687/41; and an early nineteenth-century example: 1934-98.
114 Hilaire-Pérez and Rabier, 'Self-machinery?', p. 475.
115 See, for example, *Hampshire Advertiser & Salisbury Guardian* (12 August 1837).

116 Alfred Smee, *Vision in Health and Disease: the Value of Glasses for its Restoration and the Mischief Caused by their Abuse* (London: Horn, Thornthwaite and Wood, 1847), pp. 37–38.
117 Reiser, *Medicine and the Reign of Technology*, Chs 2 and 5.
118 Smee, *Vision in Health and Disease*, pp. 40–42.
119 William Mackenzie, *A Practical Treatise on the Diseases of the Eye*, 4th edn (London: A. and G.A. Spottiswoode, 1854), pp. 915–916. For information on Mackenzie see 'William Mackenzie Medal', *British Journal of Ophthalmology*, 12.12 (1928), 648–649.
120 Smee, *Vision in Health and Disease*, pp. 45, 47.
121 Mackenzie, *A Practical Treatise*, p. 910.
122 Ibid., p. 912.
123 Ibid., pp. 987–990; cauterisation of the urethra and surrounding areas was considered a possible cure because asthenopia in some cases was linked to masturbation.
124 *The Athenaeum*, between 4 June 1853, p. 665 and 16 July 1853, p. 842; see also his later treatise: William Ackland, *Hints on Spectacles: When to Wear and How to Select Them* (London: Horne & Thornthwaite, 1866).
125 Long, *Spectacles*, pp. 23–5.
126 The original letters were destroyed in the Second World War. The surviving journal article that details the destroyed scrapbook was published in C.W. Dixey & Son, *A Short History, 1777–1977* (Newport: Mullock & Sons, 1977), p. 29.
127 Ibid.
128 Turner and Withey, 'Technologies of the body', p. 786.

2

The 'normal eye' as seen through technology: a quest for medical control, 1850–1904

A vision test was absent from early Victorian visual aid dispensing practices. Today, it plays a central role. It involves sitting in a chair and looking at a series of letters – or optotypes – at a set distance on a wall in front of you, reading what you can. If necessary, lenses are changed while you do so. The aim of the test is to ensure that you can clearly see the 20/20 standard line. The phrase '20/20' vision has become synonymous with this measurable scale and simply denotes your ability see twenty feet away what should 'normally' be seen at this distance. It is a scale of measurement most often taken for granted but has a long history; a Victorian standard measurement of capacity that was artificially created and based on characteristics of a growing and modernising society. Indeed, those who do not require spectacles often find that they are able to read beyond this standard to the two lines below. Why was a lesser degree of visual acuity chosen? What are the circumstances that led to the enshrining of 20/20 vision in medical thought and why did it become a measure of functional capacity?

The 1850s and 1860s were decisive in shifting medical understanding of spectacles and visual capacity. While medical practitioners were not, and have never been, unanimously in favour of spectacles as a method of treatment, it is possible to trace a growth in more favourable opinions from the middle of the nineteenth century. The invention of the ophthalmoscope in 1851 by German physician and physicist Hermann von Helmholtz, and its subsequent adoption and adaptation by medical practitioners, transformed medical understanding of vision. As a purely diagnostic tool, the ophthalmoscope allowed its user to observe the inside of the human eye for the first time and thereby expanded the possibilities of medical knowledge. For the growing discipline of ophthalmology, it enhanced a person's capability to determine diseases of the eye and examine the eye's visual acuity. The vision test, coupled with the introduction of new technology, identified the optical lenses a person required and simultaneously allowed practitioners to generate statistics and functional norms based on new methods of measurement. Unlike their early Victorian counterparts, ophthalmologists could now determine, explain and treat the refractive condition of a person's eye. The subsequent adoption of statistics and the privileging of an objective method of examination created an opportunity for ophthalmologists to present and assert the treatment of vision as part of the remit of medical knowledge and practice. As parallel studies have shown – for example, the emergence of otology in the nineteenth century and the debates between neurosurgeons and neurologists in the first half of the twentieth century – professionalisation, technology and competition shape, and are often crucial in, the formation of a medical specialty.[1] I do not intend in this chapter to produce an internal account of, or to explain, the emergence of ophthalmology as a medical specialty, which has been briefly outlined in Chapter One and, more thoroughly, in other works.[2] Instead, in this chapter I will demonstrate how the socioeconomic interests of ophthalmologists involved in vision testing, and the political and cultural milieus in which they were generated, shaped the attention of practitioners. These ophthalmologists were sympathetic to broader societal change and popular demands and, like later pioneers in orthopaedics, 'embraced the state'.[3] Such a position was especially important in the British context where mistrust of specialisation remained rife.[4]

In shaping their field of expertise, for the first time ophthalmologists measured visual capacity against a newly devised standard, which both offered newfound opportunities for the use of spectacles as a treatment method and subjugated vision to testing around a new and rigidly fixed norm. During the nineteenth century medical statistics became privileged in shaping contemporary understanding of bodily capacity. While scholars have taken a varied approach in defining this period of change – ascribing more narrow time frames (for example, 1820–1840) or broader notions of 'the nineteenth century' – the use of statistical tools to understand bodily norms was established in a variety of disciplines, including medicine. Lennard J. Davis has argued that the application of the mathematical norm to, and its conflation with, the natural body reconceptualised our understanding of disability.[5] More recently, however, Coreen McGuire has highlighted that the role of technology in this process has been underestimated. By exploring the technological construction of hearing loss and breathlessness, she has exposed the role of objective diagnostic tools in both gathering and shaping medical statistics and the expected standards and parameters for measuring the body during the period between the twentieth-century world wars.[6] In this chapter, I push the idea of mechanical objectivity and measurement instrumentation back into the nineteenth century by looking at the role of ophthalmoscopic measurement and diagnosis on our understanding of vision. Like Davis's argument that the invention of norms and normalcy reconstructs any deviation and the deviant body as a 'problem', I highlight how the measurement of the eye through the ophthalmoscope widened the range of eye conditions that were thought to require treatment. Similar to the use of diagnostic tools between the First and Second World Wars, medical practitioners used a new form of technology to control the methods, and determine the degree, of treatment in order to justify their increasing role and intervention in vision testing and spectacle use.

I approach a study of 'normative' vision and the standardisation of bodily measurements with trepidation. As Peter Cryle and Elizabeth Stephen's comprehensive study demonstrates, the 'normal' is often studied uncritically as an oppressive, uncontested power that allows people to be confined to deviant or pathological categories. The authors also argue that in its now mainstream usage the 'normal' is a twentieth-century phenomenon. In the nineteenth century, the normal was a scientific and not a popular or medical concept: 'there were no average patients... no standard illness'.[7] In 1852, however, vision was standardised in relation to a 'norm' and this 'norm' implied understandings of health and the performance of a person's visual acuity. Nevertheless, my work, like that of Cryle and Stephen, seeks to limit the norm's 'epistemic scope' by exploring both sides of normativity.[8] While scholarly studies of the 'norm' have often focused on its standardising function, equal attention needs to be paid to the scope it offers to differentiate and individuate. It is not, in this case, to underestimate the responses to normative vision standards, or their power over, the measurement of people's visual capacity or lived experiences, but to expose the thought processes behind the standard and the way it was conceptualised. As this chapter will show, the 'normal' and 'abnormal' was not conceived by all medical practitioners as a rigidly fixed regulatory power, and its contested meanings were often in conflict with its application. This is because treating the 'problem' of 'abnormal' vision was increasingly undertaken by medically qualified practitioners who sought new sites to test vision despite their acknowledgement, and in the face, of this conflict.

By drawing upon medical literature, medical journals and medical trade catalogues, this chapter will show how ophthalmologists attempted to normalise the expected standard of vision and determine how it would be measured both amongst the general population and in particular environments, namely schools and the workplace. In addressing the work of Cryle and Stephen I will also draw upon Georges Canguilhem's *The Normal and the Pathological* to problematise the way in which the average can be, and has been, conflated with the normal in medicine.[9] From a personal or individual perspective, Canguilhem – not too dissimilarly to Cryle and Stephen – identified that the 'normal' has no meaning because it removes the natural variabilities in a population; real people will undoubtedly diverge. Despite this, he highlights how exploring the adoption of 'norms' is a useful point of study when a norm is conceived as social and not vital or physical. He uses the example of life span and suggests that the medical finding of the 'average span is not the biologically normal, but in a sense the socially normative'.[10] Viewing the standards of vision that were devised in the nineteenth century through this lens thus reveals somewhat more about medical practitioners' expectations of nineteenth-century British society than it does about the physiological or anatomical reality of the eye. A degree of vision became socially normative and technology both measured and allowed a person to conform to this standard. As will be seen, the 'norm' in nineteenth-century Britain, Europe and America was based on a need for a person to function in a world that privileged the use of small print, signage, the need to see accurately at long distances and the need to see in increasingly confined spaces. Medical involvement and the use of diagnostic technology in vision testing redefined the theory related to vision enhancement in relation to the changing nineteenth-century world. But, as I will show, the 'normal' eye was constructed in relation to its own environment and the perceived needs of vision by the practitioners that created it, and therefore this social and anatomical 'norm' was contested and the focus of debate.

Defining the 'normal eye': the technological and medical construction of vision

The seminal work of the Dutch ophthalmologist Franciscus Cornelis Donders, *On the Anomalies of Accommodation and Refraction of the Eye*, published in 1864, transformed medical understandings of the refractive condition of the eye.[11] Donders drew upon the ophthalmoscope, recently invented by Hermann von Helmholtz, to define healthy and pathological ocular states, and determine new methodologies for testing visual acuity. The importance of the ophthalmoscope lay in the one decisive outcome of ophthalmoscopy: the ability to see inside the eye. Scholars have considered the ophthalmoscope to be the most important development of the nineteenth century for understanding the anatomical condition of the eye.[12] The Victorians, however, also acknowledged its importance in determining the eye's refractive capability and therefore the overall viability of spectacle use in medical practice. In May 1888 at the 'Presentation of the Charter of the Donders Foundation' Donders reflected on the importance of Helmholtz's 1851 invention for his own work.[13] In the past, opticians and medical practitioners understood the principles of optics and applied these to the eye in order to speculate how its shape could affect the way in which the light refracted into the retina. The ophthalmoscope allowed these theories to move beyond the realms of speculation; Donders was able to observe and determine the position of light refraction and the shape and interior of the eye for the first time. The ophthalmoscope was thus part of a much broader category of diagnostic equipment that, like the x-ray and laryngoscope, could allow its user to 'see' inside the body.[14] My inclusion of the ophthalmoscope as a central tenet of this chapter's argument creates the impression of a singular, unchanging diagnostic instrument. Such a reading would be false. Helmholtz's ophthalmoscope, as Chapter One explained, was not the first attempt to improve the objectivity of the vision testing process. Specifically, Helmholtz's instrument built upon the work of a Czech scientist and an English ophthalmic surgeon.[15] Equally, following its invention, it was not the only model of instrument utilised in the later nineteenth century to view the back of a person's eye and its use was met with resistance by some practitioners and prospective patients.[16] Indeed, Helmholtz's instrument continues to be celebrated as the 'greatest of ophthalmological instruments' but he was little involved with the variety of adaptations that took place over the remainder of the century.[17] The 'ophthalmoscope' was a multifarious category of instrument and nearly all eminent nineteenth century ophthalmologists developed new models of it – including designs that did not go into production – in order to improve its functionality. Practitioners also favoured different methods of application and could utilise the instrument in varying ways, for example, direct and indirect ophthalmoscopy. Rather than being a narrative of technological determinism, this shows the 'ophthalmoscope' to have been a fluid instrument that evolved alongside, and was influenced by, the practice of the ophthalmologists who designed and produced them.

The initial use of the ophthalmoscope by Donders expanded ophthalmologists' and opticians' knowledge from two conditions of the eye – myopia (short sight) and presbyopia (old sight) – to myopia, hypermetropia (long sight), astigmatism (a non-spherical eye) and presbyopia. Presbyopia was distinctive. Myopia, hypermetropia and astigmatism were refractive conditions of the eye resulting from malformations of the eyeball, which affected a person's visual acuity and the angle of the light ray in relation to or its distance from the retina. In contrast, presbyopia resulted from the stiffening of the eye's lens as a person aged; it was equated to the various other ageing processes of the human body including the wrinkling of the skin, stiffening of the joints and the greying of a person's hair.[18] This difference is fundamental. For the first time the various conditions of the eye were no longer grouped together as 'good' or 'weak' vision and were instead separated into two contrasting states: first, a 'natural' process of ageing and, second, an 'erroneous' refractive condition of the eye resulting from malformations of the eyeball. According to Donders, the lessening of a person's visual acuity in old age was considered to be 'the normal condition of the eye at a more advanced period of life', while other conditions of the eye, such as myopia, were based on 'abnormal construction'.[19]

To fully embed the concept of normalcy and abnormalcy in our understanding of visual acuity, Donders used broad categories to describe the conditions that he had diagnosed. He was authoritative on the use of 'emmetropia' to denote a properly constructed eyeball and 'ametropia' to denote any eye that had a refractive anomaly. How these terms related to the 'normal' and 'abnormal' eye, however, was not straightforward in Donders' text. According to the author the emmetropic eye was one that, when at rest, brought parallel rays to focus on the retina. Within this definition, Donders was quick to point out that the state of 'emmetropia' did not necessarily mean that the eye was 'normal'.[20] The term emmetropia denoted only that the parallel rays fell accurately or precisely on the retina; but this could be found in eyes with anomalies or disease. Instead, according to Donders, 'normalcy' in the eye was determined by its overall refractive and 'accommodative' visual capacity. He argued that only emmetropic eyes with correct accommodation, to allow objects to be seen equally accurately both close by and far away, could be regarded as normal. At this point the normal was not conflated with the average. Indeed, in comparing emmetropia with ametropia Donders explores whether the emmetropic or 'typically normal' eye can also be considered the most common or 'ordinary eye', and ametropia the 'exception'.[21] In this discussion, he stresses that, mathematically speaking, no eye is emmetropic because all eyes have a slight deviation and that emmetropia is not necessarily the most desirable condition because it is better for an eye to be very slightly short-sighted.

Donders did not view the terms normal and abnormal uncritically and openly questioned whether there was any relationship between what could be understood as 'normal' and the ideal or most common states of the eye. However, while his analysis of the normal and abnormal eye is nuanced and forms a less than discrete or confident conclusion, the terminology he used within his work – 'accurate', 'precise', 'within measure', 'normal' – influenced the way that the eye would be understood. Picking up from the work of Donders, for example, a British ophthalmic surgeon in the early 1860s wrote in his text that Donders classifies eyes against the 'normal or emmetropic eyes'.[22] Moreover – given the variation in the ophthalmoscope's reception – you did not need to incorporate the ophthalmoscope into your practice to be influenced by Donders' broader ideas. Later in the century other ophthalmic surgeons argued that it was 'said by Donders' that emmetropic eyes are in 'correct measure' or 'proportion' and that an emmetropic eye can be 'conveniently described as a normal eye'.[23] Perfection and imperfection, normalcy and abnormalcy, in measure and out of measure, became synonymous with emmetropic and ametropic vision. Importantly, this was against a backdrop of ophthalmic surgeons continuing to discuss and analyse Donders' conundrum: whether the emmetropic eye, as a 'state of perfection', was the most typical or desirable condition.[24]

The influence of technology and imposition of norms on vision testing was not isolated to the determination of refractive conditions; it was rooted in the way in which vision was measured. In conjunction with the development of the ophthalmoscope, a number of innovations were made in developing a method that would enable individuals to investigate and test the power of a person's vision. The most desirable of these was a system that allowed the acuteness of a person's vision to be expressed by a number. As Peter John Brownlee has argued this system relied upon developments in typography and display types which enabled a variety of European ophthalmologists to begin developing optotypes (a system of measurement that sought to achieve a formula between the size of a given letter and the distance at which it could, or should, be perceived).[25] Dr Herman Snellen's optotypes in 1862 filled a long-felt need in supplying a test that was able to measure a person's accuracy of vision on a standardised scale that would allow practitioners to compare people's vision.[26] Snellen's test coined the phrase 20:20 vision, based on the ability of a person to see letters of a certain size from a distance of twenty feet. In discussing this principle, Snellen made 20:20 vision synonymous with the 'average' and 'normal'.[27] Measuring the body's physiological functioning numerically expanded rapidly in the mid-nineteenth century but, as Stanley Reiser has argued, required contestable normal and pathological states to be devised.[28] The Victorian vision test correspondingly identified a person's visual acuity by measuring it numerically in relation to a standard. Donders did argue that Snellen's test correctly selected a 'sufficient accuracy of vision'. However, at the same time, he showed that the standardised measurement was not necessarily a reflection of reality. A study by one of Donders' pupils revealed that, earlier in life, people would exceed the standard and, later in life, visual acuity would drop by a 'normal' process of ageing.[29] Vision testing by test-types standardised the 'normal' visual acuity of humans so as to determine the degree of a person's refractive anomaly but it was not an average that fully reflected the population; it could be succeeded or exceeded. Thus, by the mid-nineteenth century, an arbitrary scale was in place for determining the degree to which the eye would be measured, compared and ultimately treated.

The measurement of the body to determine 'norms' of range, size and capacity was not unique to the testing of vision. As outlined in the introduction to this chapter, the use of statistical tools and classification systems generated and drove standardised measurements of the body in the nineteenth century. Philosopher Ian Hacking, for example, describes the years immediately preceding the invention of the ophthalmoscope, 1820–1840, as being characterised by an 'avalanche of numbers'.[30] While the concept of order and deviation was not formulated during the nineteenth century, normalcy replaced earlier words, including 'regular' or 'natural', and introduced statistical methods to the social sciences. The work of astronomer and mathematician Adolphe Quetelet (1796–1874) pioneered this approach and highlighted how human populations could be studied to generate statistics on normal distribution. Isolating individual characteristics, such as height, he showed that a population sample would cluster around an average and create a 'bell-shaped' curve. Quetelet's study and the concept of 'the average man' positioned the average as both ideal and normal. Subsequently, Lennard J. Davis has argued that this positioning had a fundamental influence on contemporary understanding of bodily capacity, deviation and disability. The average was an artificial statistic generated from the study of a population sample; the very nature of its position on a bell curve illustrates the reality of human variation. Thus, in focusing on categories of (here) deafness, Davis argued that conflating the 'normal' body with the 'average' constructed the disabled body as a new 'problem'. Any deviation was abnormal, less than ideal, and therefore could supply grounds to justify intervention.[31]

Coreen McGuire's work on deafness and breathlessness between the two world wars has extended Davis's argument and reinforced the importance of categories of normalcy for shaping contemporary understandings of bodily capacity.[32] During that period, technological instruments and statistics were crucial tools in allowing medicine to determine what historically had been, and was currently, considered to be 'normal'. Her account demonstrates how a standardised measurement is constructed, both technologically and statistically. McGuire's work highlights the need for disability history to engage with science and technology studies. Conversations between these disciplines have been hampered by disability history's politicised distancing from the medical model of disability, which has considered medical technologies as oppressive. Later social and cultural models of disability have been rightly keen to promote how society, and indeed medicine, created the problem of disability and oppressed those it identified as disabled. However, a lack of engagement with the historical processes that governed the medical model of disability has meant the use and involvement of medical technologies in categorising and conceptualising disability has not been fully interrogated. This approach has also failed, as Cryle and Stephens explain, to critique the concept of normativity and further feeds into its, unwarranted, regulatory power.[33]

McGuire demonstrates through an examination of the audiometer and the spirometer that understandings of breathlessness and hearing loss were technologically constructed in the period she examined. In particular, this understanding was grounded on the idea that instruments and machines were trustworthy and objective. This concept was not new to the 1920s and 1930s: 'mechanical objectivity' – a phrase coined by Daston and Gallison – was the yardstick of scientific study and investigation within medicine by the late nineteenth century.[34] The use of the ophthalmoscope and mechanical objectivity in formulating medical understanding of vision, vision loss and treatment in the mid-nineteenth century indicates that the technological construction of bodily capacity and/or disability has a longer history. Medical practitioners adopted the language of normalcy and abnormalcy and relied on the ophthalmoscope and an increasing number of instruments to control and define the limits and treatment of visual capacity. By prioritising objective methods, practitioners were able to reinforce their position as experts and thus attempt to legitimise their claims that vision needed to be corrected or enhanced to meet newly devised standards. Such methods were not novel. Jaipreet Virdi-Dhesi's work on nineteenth-century aurists, for example, indicates how a comparable diagnostic instrument – the cephaloscope – was used in a similar way to strengthen the field's reputation and authority.[35] Delia Gavrus's work also demonstrates how the boundaries of medical specialisms can coalesce around new technologies and, in the case of early twentieth-century neurology and neural surgery, broader therapeutic principles such as new drugs.[36] Nineteenth-century ophthalmologists' methods had two key primary effects that will be explored: first, they transformed medical perspectives on the overall utility of spectacles and, second, they urged the importance of a particular standard of vision in a range of different environments. Fundamentally, medical conflation of the average with the norm meant that practitioners' methods did not necessarily correlate with the needs and experiences of users and instead reflected medical attempts to determine a standard for the visual capacity of human populations in an increasingly visual Western world. This ambition was of interest to both them and the state, an institution with which they sought to collaborate in a new position as public experts. The application of medical technologies within medical practice was shaped by politics.[37]

Exploring the language of science adopted by practitioners is important for considering the role of mechanical objectivity in constructing nineteenth-century understandings of visual capacity and the ophthalmologist's own professional identity.[38] In 1899, the American ophthalmologist D.B. Roosa proposed that positive changes in vision testing had arisen because of medical practitioners' innovations and stated that:

> [t]he haphazard and insufficient methods of the opticians were replaced by exact and scientific measurement of the refraction and accommodation of the eye, by skilled men who could distinguish between incipient or advanced inflammatory or other morbid changes constituting disease, and purely optical conditions.[39]

Roosa's conclusion was neither new nor isolated to America. Internationally, ophthalmologists laid the importance of Donders' work with the ophthalmoscope on the newfound ability to make vision testing both 'accurate' and 'scientific'. In 1862, as that work was emerging, a British ophthalmic surgeon, Joel Soelberg Wells, had argued that methods adopted by opticians were generally 'haphazard' and 'empirical' when they could be 'scientific' and 'skillful'.[40] In contrast, following the publication of Donders' ideas, Johann Friedrich Horner, Swiss ophthalmologist and professor at the University of Zurich, celebrated the new understanding of vision testing, no longer based on guesswork and speculations that had not been thoroughly investigated.[41] Crucially, one of the perceived strengths of Donders' work was the opinion that it was based on 'scientific principles'. Similar terminology was used to describe the utility of Snellen's test-types in vision testing when they were adopted alongside the use of instruments. Both British and American ophthalmologists commended test-types for their ability to measure the degree of someone's visual acuity with 'exactness'.[42] Indeed, it was argued that the invention of test-types in itself 'virtually solved the problem of measuring and registering vision'.[43]

An increasingly 'scientific' understanding of vision and vision testing influenced medical practitioners' theories and practice. Use of objective controls to enhance the subjective examination of trial and error, as discussed in Chapter One, increased in medical texts in the latter half of the nineteenth century.[44] New methods of vision testing privileged objectivity and the ability to remove, as much as possible, any form of reliance on the patient in the diagnostic process. The method of trial and error that opticians had adopted for centuries was thus increasingly standardised through a growing consensus for a controlled testing environment and the use of standardised testing equipment. In 1864, Donders pinned the development of the objective examination of the eye on the ophthalmoscope. This did not lead to complete dismissal of the utility of trial lenses and test-types for determining simple refractive vision errors. Rather, the Donders method combined objective and subjective methods by utilising trial lenses alongside test-types and the ophthalmoscope to determine a person's visual acuity.

A greater understanding of refractive vision errors, vision testing and diagnostic equipment changed the medical stance on spectacles. The use of technology not only generated the 'problem' of substandard vision, but also encouraged the use of technology in 'correcting' it and ensuring that people could conform. Donders' treatment of the condition of asthenopia illustrates the influence of his work on the adoption of spectacles as a treatment method. He revealed that 'asthenopia' – a condition of the eye discussed in Chapter One that causes a range of debilitating symptoms such as headaches and physical weakness – was often caused by hypermetropia, which he had diagnosed for the first time in 1858.[45] In most cases, Donders was able to put forward evidence that understanding and measuring hypermetropia, and then treating the condition with spectacles, successfully remedied a fatiguing condition that could limit the quality of life. He questioned why previous examinations had not discovered the connection and concluded that 'it is a great satisfaction to be able to say that asthenopia need no longer be an inconvenience to anyone'.[46] The quick transmission of Donders' ideas in Britain can be seen in British ophthalmologists' work.[47] Donders defined a refractive condition, disproved earlier theories and showcased the effectiveness and benefits of treating refractive vision errors.

The success of his treatment of asthenopia was based on the adoption of new technologies and a more thorough examination. Ophthalmologists promoted the incorporation of a variety of equipment to increase the objectivity of the examination, including viscometers, retinoscopes, optometers and ophthalmometers. Additionally, these instruments, and particularly the ophthalmoscope, were developed and improved regularly across the second half of the nineteenth century. Medical trade catalogues are significant for showing doctors' interaction with traders of instruments and material goods.[48] The ophthalmology section of medical trade catalogues illustrates the expansion of medical practitioners' diagnostic equipment. In 1863, for example, London-based James Weiss & Son stocked one type of ophthalmoscope, and a trial lens case. By the 1880s and into the 1890s they greatly expanded their range.[49] This pattern existed in catalogues from both municipal and provincial suppliers, and expansion in stock typically extended to an increase in the types of ophthalmoscopes, test-types and trial lens cases on offer.[50] Consecutive runs of catalogues for individual companies highlight that the 'surgical instruments and appliances' marketed to ophthalmologists grew to accommodate the need to test vision with a variety of equipment as well as their more traditional role of treating diseases of the eye.[51]

The ability of this equipment to enable practitioners to assess vision against standardised tests of visual capacity, and treat it, was further enhanced by the adoption of the dioptre in 1875. The dioptre replaced earlier arbitrary lens measurements, which had differed amongst opticians and lens grinders and had marred any attempt to draw comparisons between different individuals' visual capacity and treatment requirements.[52] It also expanded the range of lens types from which examiners could choose by increasing the gradations between different lenses. The standardised and more finely ground lens, coupled with the use of measuring equipment, were used by practitioners to validate their claims that they could diagnose a greater number of eye conditions – such as astigmatism or hypermetropia – with greater speed and precision. It is important to note that this was an incomplete process; the dioptre was not universally used at first, and practitioners developed their own techniques for incorporating instruments into the diagnostic process. Nevertheless, the ideology behind these developments and their application was built on a general consensus. As illustrated in correspondence published in the *British Medical Journal* in 1884 by Henry Juler – aptly entitled 'the best methods of diagnosing and correcting errors of refraction' – objectivity, accuracy and reliability were the ophthalmologist's goals.[53] Juler acknowledged that trial glasses used to present a patient with lenses of different gradations were 'perfectly safe' for simple refractive errors where vision could be immediately improved to the 6/6 (or 20/20) standard. However, crucially, he concluded that relying on the patient raised a number of potential problems and the use of objective methods saved the examiner great time and trouble. Juler favoured the shadow test – a method that used a retinoscope – for determining more complex errors of refraction because it eliminated any reliance on the patient. Other practitioners' expressed preference for this method, or the use of other instruments, during the 1880s because, by this point, 'complete independence' from the patient was seen as desirable.[54] This is not to say that medical expertise and objective testing were not beneficial for prospective spectacle users. In contrast, they greatly enhanced the utility of spectacles for a range of refractive conditions of the eye. Nevertheless, despite these benefits, the new approach removed the user from the testing process; the conditions and environments in which vision was tested became increasingly controlled and new standards were subsequently developed to cover a range of different environments. The objective technologies were governed by subjective interpretation and produced subjective results.

Asserting normalcy and medical control

Ophthalmologists' writings and attempts to gain an authoritative position on vision testing and vision enhancement were part of their broader aim of achieving professional status. Medical practitioners had used the cultural importance of vision, coupled with a growing acknowledgement of its fallibility and vulnerability to disease, to successfully establish the first medical speciality that was not associated with quackery at the beginning of the nineteenth century.[55] An international community of ophthalmologists and a variety of new medical texts had emerged following the establishment of specialist eye institutions and associated training, as we examined in Chapter One. In Britain, the Ophthalmology Society was set up in 1880. In 1881, the president of the Ophthalmology Section at the British Medical Association reflected on the recent International Congress of Ophthalmology and how the position of the discipline had changed:

Our branch of practice... stood in a sort of isolation... To-day we are obviously one among many members of the entire medical commonalty, and our work is admitted to be of the highest value, not only for its own sake, but for the beneficial influence which its aims and methods exercise over the whole field of medicine.[56]

Thus, while ophthalmologists developed the usual markers of professionalism – they started their own journals, produced their own medical texts and held their own congresses – their interaction with the broader discipline of medicine and the public was perceived to be integral to expanding their influence and position. Ophthalmologists used their growing sense of professional identity to consolidate their identification of standards and share their growing knowledge of vision with medical practitioners. Jaipreet Virdi, in her assessment of the medicalisation of deafness, emphasises the importance of the Royal Dispensary for Diseases of the Ear for 'legitimising' the 'medical infrastructure' and medical control over Victorians.[57] While ophthalmologists' institutions – the eye hospitals that had sprung up across Britain – were important, by the 1880s ophthalmologists had a much broader vision. On a theoretical level they attempted to consolidate their expertise through texts that have been explored in this chapter and contributions to the *British Medical Journal* which advised on vision testing and refractive conditions of the eye. On a practical level, however, they embraced the state and drew upon widely felt concerns about deteriorating vision. Ophthalmologists increasingly sought to monopolise vision testing, sat on advisory committees, produced governmental reports and had their views published in a range of newspapers and periodicals aimed at the general public. Technology and the role of normalcy in creating a 'problem' that ophthalmologists were able to solve were integral to their discourse. Both schools and the workplace emerged as international sites of investigation by ophthalmologists and provide the best examples for exploring how the profession asserted its position and the implications of technological intervention and the categories of 'normalcy' and 'abnormalcy' for nineteenth-century understandings of vision. Scholars have illustrated the value of studying medical journals alongside medical texts in their investigation of school hygiene and working conditions in the nineteenth century.[58] However, no study of eyesight and education in a British or international context has looked beyond the medical profession to assess the influence of their ideas. This section, therefore, draws upon a broader range of medical journals and medical texts, while also exploring how nineteenth-century newspapers and periodicals presented and discussed education and work.

The school in Britain, following the 1870 Education Act, became an area of general concern amongst medical practitioners. These concerns can be grouped with practitioners' broader preoccupation with 'civilisation', namely the effects of the environment and lifestyles of those in the Western world on both visual development and physical health. Steps to make elementary education compulsory led to debates over its long-term impact on children's vision. In 1887, for example, Dr Arthur Newsholme, who wrote extensively on issues of public health, stated that myopia 'may be acquired, and it is chiefly during school-life that this occurs'.[59] With growing concerns in the nineteenth century about national degeneration and poor physical health, children were considered representative of the future nation's wellbeing. A number of school hygiene texts by general medical practitioners emerged, and in these vision appeared prominently. In popular and medical discussion schools' deficiencies were both material – desk and lighting – and practical – type and length of study.[60] However, ophthalmologists were increasingly part of these textual enquiries, produced their own bodies of work and held a prominent position in broader debates in the *British Medical Journal*, as well as interacting and engaging with international studies done by ophthalmologists in Europe and America. Richard Meckel has argued that American investigations into schools and the implication of school conditions on children's health pathologised the body and created a host of new norms.[61] Similarly, in ensuring that vision was protected in the young, British ophthalmologists both normalised and pathologised children's visual development.[62] In turn, this created and established the measurement of vision as an increasing part of the discussion and role of ophthalmologists and the medical profession.

A proliferation of articles on the condition of children's eyes emerged in the *British Medical Journal* alongside a host of ophthalmologists' texts devoted to lectures on the subject from the 1880s as ophthalmologists consolidated their position as a distinctive and acknowledged medical specialty. The primary purpose of these publications was to draw comparisons between international studies of significance and provide new statistics on the relationship between education and eyesight in children. Medical interest in the topic is reflected in correspondence in the *British Medical Journal* which, for example, featured the topic of 'Eyesight and the Education Act' in 1889.[63] However, published lectures from various prominent British ophthalmologists indicate the reach and influence of medical ideas on understandings of the relationship between schools and vision. Three key lectures that had been delivered to educational organisations were published independently between 1880 and 1885. In November 1880, for example, a Birmingham ophthalmic surgeon, Priestley Smith, gave a lecture to the Birmingham Teachers' Association.[64] Similarly, Simeon Snell, ophthalmic surgeon to the Sheffield General Infirmary, delivered a lecture to the Sheffield and District Certificated Teachers' Association in January 1884.[65] On a national level, Robert Brudenell Carter, ophthalmologist and Hunterian Professor of Pathology and Surgery at the Royal College of Surgeons of England, delivered a paper to the Association of Medical Officers of Schools in April 1885.[66] As indicated by the involvement of medical officers, the investigation into the matter and the publication of their findings was part of broader governmental enquiries.[67] These studies proposed that while current education was overworking and straining students' eyes, it was not study itself that was causing vision defects, but the poor lighting and desks with which children studied, both at schools and when completing homework.[68]

Jamie L. Bronstein has used the proliferation of newspaper articles on workplace accidents in the nineteenth century as evidence that readers focused intensely on this issue.[69] The large quantities of information in both medical and popular texts suggest that the fear and anxiety about nineteenth-century education, and its association with the deterioration of eyesight, was both a medical and a public concern. Moreover, it highlights the position of medical men in controlling and determining the level and nature of this concern. British newspapers and periodicals provided information on specialised ophthalmological works and commented on investigations that were attempting to generate new statistics on children's vision. Current studies, for example, were reviewed and Carter's books, governmental reports and academic papers were particularly prominent in the 1880s and 1890s.[70] Likewise, the findings of Snell and Smith were influential and reviewed in popular forums.[71] Reports on lectures revealed interest in localised studies that were undertaken by ophthalmologists. In 1890 a paper was reported in the *Aberdeen Weekly Journal* that, for instance, included some notes from an investigation into children attending Aberdeen public schools.[72] In 1899 another report on eyesight from schools in the local area, which had been requested by the government, appeared in the *Leicester Chronicle*, and this illustrates that its findings were considered of both governmental and local popular importance.[73]

International works of significance were taken up in Britain in response to this increasing national interest and a landmark work on the school environment by German ophthalmologist Richard Liebreich was translated into English in 1878.[74] Strikingly, however, the influence of the professionalisation and increasing establishment of ophthalmology for public understanding of schools and vision is best seen in the way in which international studies were reported. Although a British ophthalmologist, James Ware, was credited with pioneering the study of vision and education in the early nineteenth century, little work had been done in Britain between 1814 and 1880.[75] Indeed, many of the ophthalmologists writing in the 1880s commented on the lack of comparable study of vision and short sight in Britain. American medical journals and texts reveal that the subject was given attention a decade or more earlier there than in Britain.[76] In Europe, Britain lagged more severely behind. British studies followed broader trends in the professionalisation of ophthalmology and understanding of refractive vision errors in Germany, in particular, which had undertaken a number of significant studies as early as the 1860s. Moreover, these German studies were neither published nor gained popular interest in Britain until the 1880s. The pioneering examinations by German ophthalmologist Hermann Cohn between 1864 and 1865, for example, were repeatedly published from 1880 in British popular print. Interest in Germany was based on its status as having the worst eyesight in the Western world, which was attributed to both conditions in schools and German type and printing techniques.[77] However, interest in Germany in the 1880s was also facilitated by British ophthalmologists' increasing prominence and bringing the issue into public light. Further findings were also reported from studies in France and America, which detailed how they were attempting to halt the acceleration of vision defects through school design.[78] Significantly, while the reports on these new school design initiatives were written as the events occurred, it was only when works by British ophthalmologists began to appear in the later nineteenth century that earlier work by authors such as Cohn began to receive comment in British newspapers and journals.

Interest in the international context occurred in tandem with the uptake of the subject and intervention of ophthalmologists in British schools and the state of British children's eyes. Ophthalmologists acted upon this trend and acknowledged the role of popular texts in enabling a broader proportion of people to access large quantities of referenced information. In a bid to encourage change, for example, they could write explicitly for a pedagogical audience. However, ophthalmologists also had to compete with public health experts. In 1900 Dr Newsholme, for example, wrote a regular column in multiple editions of the *Practical Teacher*. The column intended to educate teachers on medical ideas. In a discussion of homework, for instance, Newsholme utilised the same information that he had presented in his own more medical-focused treatise.[79] Similarities in the form and content of text produced highlights how the boundaries between readership and different types of text are not easy to demarcate. It would be too simplistic to conclude that the general public read popular texts, or that medical texts were only consulted by a medical audience. For instance, Carter's text on vision and eyesight was read by a wider, less educated, audience.[80] Indeed, both the publication of medical books and reporting of their findings in newspapers and periodicals served to assert medical influence. Yet, despite the similarities, the differences in style could be more explicit. The language adopted in certain newspapers and periodicals highlights that ophthalmologists' views and the position of vision testing were gaining attention and being distorted. At the outset of his 1885 study on schools and vision Carter, for example, was doubtful of the influence of schools and declared 'nothing is farther from my own wish than to play the part of an alarmist'.[81]

In contrast, he criticised a newspaper that had reported on one of his previous studies and selected certain evidence and items of content to scaremonger.[82]

While the distortion of evidence was criticised, newspapers and periodicals played a vital role in drawing popular attention to the 'problem' of vision and the role of medicine in intervening and providing a solution. Vision was one of many areas in which investigations into schools and education in a broader international context led to standards and norms for children's development. Moreover, Chris Otter has argued that British investigations into the environment and practice of education in schools attempted to 'normalise' the projected visual development of children. He proposes that, although slow and uneven, the adoption of vision testing was 'discernible' by the end of the nineteenth century.[83] Similarly, J.B. Hirst has argued that Carter's 1896 governmental inquiry led to both teachers and ophthalmologists taking a role in the testing and measurement of children's vision in schools.[84] Carter had focused since the 1880s on the responsibility of schools and the state to ensure children's eyes were tested prior to the outset of education. This recommendation filtered into newspapers and periodicals for the remainder of the century. Who should provide vision testing emerged as a topic of debate in *The Standard* in the 1880s, which reflected on the repercussions for the 'poor little sufferers' who were not tested.[85] Within these discussions, responsibility could be placed on the medical practitioner, the teacher and also the parent. In 1892, for example, an article in *The Yorkshire Herald* (quoting from *The Lady*) provided popular advice and suggested that parents with visual defects had an enhanced responsibility to get their children tested. It concluded that 'the remedy for all this is thorough and careful and regular periodical examination'.[86] However, while influential in its use of standardised technologies and diagnostic categories of normalcy and deviance, the implementation of vision testing was very much in its infancy in the 1880s and 1890s. Concern about the extent and effectiveness of testing initiatives was reflected in the *British Medical Journal* up to the turn of the century and discussion on school education continued to call for three steps that revolved around improved, more regular, vision testing.[87]

There are no statistics on attendance for vision testing in the 1880s and 1890s, to ascertain how far medical ideas were taken up in practice. Despite this, popular discussion of medical ideas surrounding vision and education made a distinct contribution to understandings of vision and how children's vision should be both investigated and categorised. Medical ideas framed a child's vision as something that could be both measured and standardised, as well as increasingly subject to deterioration. The strain of education on children's vision was explained by many British and American ophthalmologists to be the result of close work, and in the context of broader, systemic views of the body. Homework, for example, was considered particularly burdensome because it interrupted hours of 'ease' or 'play'. In these discussions, exercise and the physical health of children were highly influential. Vision defects were therefore particularly alarming to the medical elite because they were believed to be rooted in the broader condition of the body and came to be associated with physical weakness and poor mental aptitude. The toll of refractive vision errors was thought to have a marked effect on a child's physical appearance. Those with myopia, for example, were described as 'stooping' and becoming 'quiet' and 'pale'.[88] This reference to pallor was part of a wider association between short-sightedness and unhealthiness in medical texts and served to reinforce the 'problem' of vision that medical practitioners both measured and sought to 'correct'. Moreover, stooping was associated with the influence of vision on the body's broader physical frame. There was consistent discussion of the malleable effects of vision defects and the subsequent development of scoliosis. Practitioners described children affected by these defects using a range of alternative terminologies that implied physical weakness, such as 'delicate'.[89] It was continually emphasised that the myopic eye was not a 'strong' eye and this was frequently placed in direct relation to the body's constitution:

> A sound physique – a healthy vigorous condition of the body generally – is antagonistic to the working of almost every morbid process, and even in the case of short sight... I will merely point out in passing that an impaired physique brings with it, amongst other evils, an impaired resistance to the active causes of short-sight.[90]

Yet, while medical discussion of vision's deterioration and its effect on children's health raised alarm, such discussion also brought medical assertions under scrutiny. In 1896, for example, an article in *The Morning Post* postulated that 'there is an obvious difficulty in apportioning such an increase between a more general recognition of defective eyesight among children and a growth of the defect itself'.[91] The article did not question medical measurement and diagnosis of the eye, but did propose that the acceleration in vision defects being reported among children in the nineteenth century may have been because of increased attention to the issue rather than an increase in actual numbers. Peter Conrad, in his 2007 study of the medicalisation of society, posed a similar question that can be applied more broadly to the investigations into vision errors by ophthalmologists: 'does it mean there's a new epidemic of medical problems or that medicine is better able to identify and treat already existing problems?'[92] The Victorian medical elite, on the basis of this concern, did interrogate the extent to which this rise in case numbers and the diagnosis of children's vision as defective, weak and unhealthy was appropriate. In 1899 D. Love, for example, argued that ophthalmic surgeons were erroneous because they were 'adopting an arbitrary standard, and calling it the normal for children's eyes... Nature has no fixed standard. Here normal is a variety.'[93] Central to Love's criticism was the adoption of an artificial 'standard', which in turn led to the diagnosis of 'normal' and 'abnormal' vision. If the ophthalmoscope and related vision testing equipment were conceived as 'objective' tools, the interpretation of the findings and data gleaned from using these instruments was not.[94] Abnormal vision had become a subjective, loaded term; it was synonymous with overall bodily weakness and defectiveness, and demanded treatment. Such a term was contested, conflictual and had the potential to be misapplied.

Lighting and seating conditions in the workplace underwent similar scrutiny to the school environment, in terms of the possibility they might adversely affect people's visual capacity. However by the close of the nineteenth century ophthalmologists were less inclined to place the blame on the environmental conditions of the workplace, unlike schools, and instead focused on the worker's body and fitness for the conditions in which they worked.

Moreover, the implications of 'normal' categories of vision being applied to Britain's workforce emerge in the form of qualification and disqualification from occupational environments. Investigation of the workplace has a much longer history and therefore the trends in, and influence of, increasing medical intervention and medical diagnosis in exposing a new form of problem are more visible. It is possible to trace a clear shift from adapting and controlling environmental conditions to intervening and determining a person's visual capacity or capability. Technology features prominently in this process of identifying the 'problem' of vision and allowing medical practitioners to perform a 'gate keeping' role in determining who and who was not fit to work. As noted by scholars, the increasing measurement of the efficiency and capacity of workers' bodies led to the determination that some bodies should be excluded while others could continue work but would need enhancement. Vision held a prominent position in these debates because of the elevated role given to a measurable level of visual acuity, which helped to coin phrases such as the 'economy of the eyes'.[95] When measuring vision, inclusion or exclusion was based on the adopted, but subjective, 'normal' standard.

Occupational environments became an issue of public concern in the Victorian period before ophthalmology fully established itself as a discipline. The mid-century, galvanised by attempts since the late eighteenth century, saw a wealth of legislation that sought to impose regulation on the industrial workplace. Moreover, the social conditions that stemmed from industrialisation helped prompt the 'Condition of England' question, which criticised class disparity and the poor domestic and working conditions of the lower classes. Social commentary, legislation and the establishment and enforcement of factory inspectors following the Factory Act of 1833 placed the workplace under greater scrutiny. Industrial accidents and workers' health were at the forefront of these concerns. Indeed, increase in workplace accidents led to greater information and discussion of the work environment both in the press and as part of governmental studies.[96] However, concerns about vision in the workplace environment extended beyond workplace accidents; it formed part of a wider interest in the effects of overuse or strain on the eyes. While Bronstein has shown that a 'health and safety' concept was not yet in place, contemporary interest reveals that there was concern over the use of eyes in the workplace, and some attempts to reduce occupational hazards.[97]

Assessing the changing influence and involvement of medical practitioners in discussion of the workplace environment is best illustrated through reports in newspapers and periodicals. Early methods of intervention by both ophthalmologists and a variety of groups focused on improving the conditions of the workplace environment so as to mitigate against deleterious effects to the eyes. In 1833, for example, *The New Monthly Magazine and Literary Journal* recommended its readers to consult John Harrison Curtis's treatise on the physiology and diseases of the eye. Curtis, part of the nascent group of oculists and aurists advertising in popular publications, was not an ophthalmologist and had not received conventional medical training.[98] Nevertheless, the periodical promoted his work and argued that the chapter on preserving the sight should be 'engraven on the palms' of those who were engaged in literary employments, such as barristers, clergymen and publishers.[99] In addition, a series of articles in *The Saturday Magazine* in 1838 on 'Employments which injure the eyesight' drew upon Charles Brewster's *Optics*.[100] These texts placed the health and strength of a person's vision in direct relation to their work environment and the primary focus was on the adverse effects of close work and poor working conditions. This accelerated by the mid-century where references to eye strain, work and eyesight in popular texts increasingly referred to the latest work by ophthalmic surgeons. In 1856, for example, *The North British Review* incorporated recent publications by Alfred Smee, surgeon to the Central London Ophthalmic Hospital.[101] Two years later, in 1858, an article headed 'A plea for the eyes' published a study conducted by the Society of Arts which had appointed a 'Committee on Industrial Pathology on Trades which affect the Eyes'. The committee did involve and include a doctor, but sought advice from surgeons who were prominent in developing ophthalmology as a discipline, such as William White Cooper, surgeon to St Mary's Hospital. The focus of medical advice or publications at this point also remained rooted in the workplace environment. The principal line of inquiry for the Society of Arts Committee was to discover what 'eye maladies' resulted from particular trades, and whether these maladies stemmed from 'injudicious management' or the 'exigencies of the worker's employment'.[102]

Within both non-medical and early medical studies, therefore, there was an early attempt to associate eye defects and conditions with certain environments and determine ways to avoid these. In 1872 this approach was made explicit at the International Oculist Meeting where statistics were given on the number of cataracts that were found in a variety of different occupations.[103] In particular, the bending or stooping postures and poor lighting conditions of artisans, lace makers, watchmakers and engravers were considered injurious, and parallels can be drawn between medical discussion of school hygiene. This association held cultural currency and it was frequently highlighted that those engaged in close work had short sight, while shepherds, sailors and farm labourers typically adapted to the conditions of their work and had long sight.[104] However, in the closing decades of the century, medical practitioners interrogated this association to assert current medical practice. A number of international medical investigations provided statistics to suggest that vision was not always marred by jobs that involved close work and that it might not be close work itself that was the problem. In 1878 American ophthalmologist Edward Loring, for example, drew upon both his own work and international studies to conclude that watchmakers and silversmiths did not suffer any injurious effects because of their work. This proposition challenged popular misconceptions in Britain. In the 1880s, for example, Carter was also a particularly strong advocate against the idea that watchmakers derived poor eyesight from close work, a claim that had been frequently printed in newspapers and periodicals in the early 1880s.[105] International studies were also considered topics of interest and, for example, an article in *The North-Eastern Daily Gazette* in 1891 referred to studies by a German ophthalmologist who had obtained similar results.[106]

At the close of the nineteenth century, J.T. Arlidge conducted an innovative and extensive assessment of the health of employees in different occupations. By drawing upon findings from earlier investigations that had been published throughout the century, Arlidge's study revealed the shift in medical influence on the use of the eyes in the workplace. Some of the older notions prevailed. Close work employments, such as lace making, artificial flower making and hosiery manufacture, for example, were associated with poor eyesight. However, by drawing upon the work of Lloyd Owen, senior surgeon to the Birmingham Eye Hospital, Arlidge revealed that watchmakers, jewellers and engravers usually, in fact, had good eyesight. This shift in medical understanding can be traced to the embedding of normalcy, disease and strength in medical understandings of eye health. Loring, for example, acknowledged that the environmental conditions of close work were not damaging in adult life, in comparison to school life, because the adult eye was 'strong enough to withstand the strain'.[107] Owen also emphasised the importance of strength. As senior surgeon he, together with his colleagues, saw an average of 20,000 cases per annum. His findings revealed that 'close work even when regular and persistent, does little, if any harm, to normal eyes'.[108] Here, it was stated that close work was not harmful to normal eyes, because it affected only eyes with latent or manifest refractive conditions. In other words, Owen concluded 'it is not the work but the unfitness of the eyes for the work which is to blame'.[109] He shifted the problem away from lighting, desk conditions and long working hours to individuals themselves. Such conclusions were part of a much broader ideology that governed the unprecedented attempt to measure the physical condition of the individual worker and their fitness for the conditions of the workplace.

In 1887 Irish ophthalmologist Dr Arthur Benson similarly based the suitability of a person for different professions on the condition of their eyesight:

> Before deciding on a profession, employment, trade, or form of labour to which any young person should be put it would be very advisable to weigh with due care the question of his sight, and the probable influence of the employment on his eyes.[110]

Benson proposed that for those with more severe degrees of myopia, or short sight, 'a country life with agricultural work was alone suitable'.[111] Ophthalmologists became more concerned with measuring the physical capacity of the worker than questioning the environment and poor working conditions. While medical understanding of the eye and the measurement of vision were important to this, this shift arose also because of two broader historical trends. First, there was a much wider interest in workers' bodies in industrialising countries. Scholars have shown how contemporaries saw labour as integral to the growth of national economies and, therefore, the health of workers and their suitability for certain jobs meant workers' bodies became crucial to the success of production.[112] More recently, Sarah F. Rose has highlighted how, in America, this increased workplace scrutiny, coupled with mechanisation and the concept of efficiency, led to the measurement and exclusion of bodies from certain workplace environments.[113] I take this argument a step further by considering a second historical trend, which is that the medical profession increasingly performed 'gate-keeping' roles to judge a person's fitness for employment.[114] Previous studies have highlighted the invasive nature of their role and have referred to general practitioners' conduct as 'medical policing' to ensure a person met certain height, weight and visual requirements for occupations within organisations such as the Post Office.[115] However, studies have not considered these findings in light of their implications for the conceptualising and shaping of capacity. These interventions can be seen as the culmination of work undertaken by medical practitioners in the nineteenth century to cement themselves as authoritative experts in matters of occupational health and the control of workers' bodies.[116]

Ophthalmologists utilised their medical understanding of the eye to regulate and ensure efficiency and safety in the workplace. In doing this, it was argued that certain occupations demanded a standardised level of visual acuity that could be measured and would determine an individual's fitness for the work required.[117] As might be expected, physical requirements for the army and related military services were heavily discussed and there was a popular fascination with soldiers who were allowed to wear spectacles in some countries.[118] Beyond popular discussion, however, medical authorities became increasingly more involved in workplace testing and vision testing featured prominently in medical journals. Articles which detailed both the physical and visual qualifications required for admission to the medical arms of wider military services, for example, began to appear in the *British Medical Journal* from the 1880s.[119] In 1885 Surgeon-General T. Longmore in his *Optical Manual* justified this intervention by arguing that the development of long-range firing increased the military importance of good sight and therefore more complex examination was required.[120] He provided a guide for a more thorough, and objective, eye examination, which in turn provided more sophisticated rules and measurements required for physical qualification. These new rules and regulations were set out for both medical officers and those on the front line. In 1895 Kenneth Macleod discussed the physical examinations required for employment in government, and related services, and the decisive new role of medical authorities. He concluded that vision testing and visual requirements for occupations had 'become an important part of medical practice, constituting boards and committees acting under rules and regulations'.[121]

As well as the civil and military services, the medical profession increasingly sought to judge a person's fitness for a variety of other employments. In 1881 E. Warlomont, a Belgian medical practitioner writing in the *British Medical Journal*, argued that an objective instrument – the optometer that was discussed in Chapter One – should be used for the examination of soldiers and railway workmen.[122] For railway employees, he considered the 'concomitant examination of the refraction and of the acuteness of sight' to be 'indispensable'.[123] Systematic vision and physical testing of railway workers and those who worked at sea became a particular legal and medical concern in the later nineteenth century.[124] Colour perception and a person's long-distance visual acuity emerged as vital with the advent of colour signalling and faster locomotion. However, despite being acknowledged as a potential problem from as early as 1850, little knowledge had been advanced in Britain. In 1881 Charles Roberts, for example, introduced the topic as a new area of inquiry for the Anthropometric Committee of the British Association for the Advancement of Science based on 'extensive investigations' in America and Europe and a lack of comparable study in Britain. He included instructions and testing materials to equip committee members with the expertise perceived necessary.[125] In contrast, medical practitioners, uninterested in gathering statistics of incidence, increasingly emphasised and turned their focus to the need for accurate examinations – to be undertaken only by those in the profession – during the last two decades of the century. In 1890 Thomas Bickerton, oculist for the Liverpool Royal Infirmary, condemned the Board of Trade for their failure to provide proper regulations (and continued to do so in 1895).[126] Bickerton, and those who responded to his condemnation in the 1890s, contrasted the Board's inaction with medical action to secure proper regulation.[127] Medical attempts to regulate workers' capacity could be seen to be motivated by, and protecting, the employer's interest. In Britain the changing legal context of the workplace, and government intervention, influenced employers' perceptions of their employees' bodies. The 1880 Employers' Liability Act and the 1897 Workmen's Compensation Act made employers increasingly responsible for safety at work. In coal mining, for example, this led to the aged or impaired being dismissed from work because they were considered more vulnerable and prone to accidents that could injure themselves and/or others.[128] However, for vision testing on the railway or at sea, medical attempts to assert new standards of 'normalcy' and 'perfection' to improve public safety were in opposition to the conduct and beliefs of private companies.

Medical practitioners argued that more sophisticated and systematic testing needed to be carried out by fully qualified individuals in the rail and marine services but that ophthalmologists had struggled to implement any form of change; standards of normalcy were resisted in Britain while Europe accepted change more readily. In 1891 a group of medical practitioners, as part of 'A discussion of the vision of railway servants' in the *British Medical Journal*, identified deficiencies in the current system.[129] Drawn from the British Medical Association's annual meeting, the discussion pointed to the inaccuracy of railway tests and the need for ophthalmologists to undertake the testing themselves, to prevent fraudulent results. It closed by reporting on an international regulation that had been formulated in 1881 at the International Medical Congress of London. This regulation stipulated that engine drivers needed to have 'normal' visual acuity and any other person employed on the railway should have 'normal' vision in at least one eye. Although ten years had passed, the British Ophthalmological Society committee had yet to address and implement the regulation. Demands for greater medical involvement were in response to opposition to vision testing from both employers and those employed on the railway. Railway workers' opposition was directed towards the invasiveness of medical intervention and its implication for people's employability. At first glance, this could be seen as an example of medical practitioners failing to understand the needs of their patients. However, vision testing was deemed a necessary development in line with international guidelines. Medical practitioners therefore believed that failing to implement it in a timely manner would be more detrimental because workers might still enter training or years of service, and then be excluded. The effect that late disqualification from service had on employees' welfare, and on health and safety, were at the centre of criticism directed towards railway companies. Two case studies, for example, revealed that men in their mid-thirties had passed a range of earlier vision tests but failed a more rigorous examination, and as a result been disqualified for the services that they had spent time training for.[130] Further evidence of individuals slipping through the examination process also appeared in the discussion of two men from the railway profession who had managed to pass five previous vision tests.[131] Medical practitioners' calls for intervention may therefore have had a less self-interested and more altruistic aim. As Jordanna Bailkin has noted, workers could 'mysteriously' be found to be colour blind and were concerned that employers, testing their own employees, could manipulate the system.[132]

With the health and safety of the public and employees in mind, correspondence in the *British Medical Journal* criticised the haphazard measures currently adopted by owners of railway companies who did not judge the risks associated with visual acuity in a similar manner to age or impairment in other industrial workspaces. As with schools, practitioners argued that Britain was lagging behind other parts of the world. Despite the implementation of the Workmen's Compensation Act in 1897, William Beaumont, surgeon to the Bath Eye Infirmary, discussed reasons for Britain's seeming slowness to incorporate standardised vision testing for employees working on the railway.[133] He first outlined the laws passed in Europe without opposition and then targeted railway owners' preoccupation with money, and their workers' dismissal of the dangers of impaired visual acuity, and both sides' opposition to the implementation of stricter measures. In attempts to assert their standards of 'normalcy' and 'perfection' Beaumont argued that ophthalmic surgeons had produced material for both medical and lay papers and published lectures by recognised authorities in an attempt to turn public opinion. While not successful in legislative terms, the public did voice their concern about safety and lack of accurate testing in several popular newspapers, which often peaked in response to railway collisions. In the *Daily News* in 1882, for example, a correspondent argued that there had 'long been an uneasy suspicion in the public mind that engine-drivers do occasionally run through danger signals'.[134] The railway and marine professions show how improvements in technology, namely the invention of coloured signal lighting and faster locomotives, created a new environment. This environment required a person's eyes to be used in new ways and served to expose a person's visual capacity. These work environments both highlighted employees' visual acuity and generated attempts to measure this against a 'standard' in a similar manner to schools. However, the competing interests of railway companies, workers, the public and medical practitioners highlight the contestable nature of the standard being enforced. While European legislation, and British ophthalmologists, stipulated that 'normality' or 'perfection' in at least one eye was required, Britain's railway owners considered that a 'man who has one quarter of the normal eye in his better eye is considered to be visually fit'.[135]

Insufficient vision testing failed the railway companies, and threatened the safety of travellers and the livelihoods of the men involved.[136] However, it also served to demonstrate the capability of medical expertise to tighten, determine and shape the bodily capacity of a workforce. At this point, testing was in its infancy and far from certain, but its long-term implications are apparent in the use of appropriate and pioneering vision testing methods in the late 1890s by some sea and railway companies. In 1895 the President of the Board of Trade drew attention to the fact that 'great lines had set a good example; the Cunard, the White Star etc., were subjecting their men to very careful tests'.[137] As proposed in 1881, the White Star Company, like the Cunard Line and the Honourable Trinity Company, had adopted periodic testing for all men employed.[138] While the extent of vision testing in these professions was not deemed satisfactory, it does not detract from the level of emphasis on the nature and importance of vision testing in shaping a measurable standard of vision as a requirement for working in certain occupations, one that did not need to be in line with the employer's or employee's interest. Under the guise of health and safety, medical practitioners were able to assert their standards of normalcy and use of more sophisticated methods of vision testing to both qualify and disqualify a worker from employment.

Conclusion

Medical practitioners constructed normalcy and normal vision in the Victorian period in relation to the environment and the perceived needs of the time. These needs were to be able to determine and compare a person's vision in relation to a standard in order to provide consistent treatment; self-interested attempts to assert practitioners' position on vision testing theory and practice; and a response to broader nineteenth-century changes demanding that vision needed to be measured and of a minimum level. In the measurement of the body and identifying the 'norm', technology was integral. Instruments offered an objectivity that was associated with 'science' and 'exactness', which changed the medical stance on how the eye worked and served to emphasise the utility of spectacles. However, it also limited reliance on both the patient and others to administer a vision test; practitioners deemed that only the medically qualified could determine whether intervention was required and conduct a vision test to the appropriate 'accuracy'. The use of objective tools was governed by subjective interpretation and had subjective results.

Ophthalmologists were pervasive in asserting and applying new medical theory and practice in a variety of changing nineteenth-century environments to develop a new type of 'modern' vision. This demonstrates on one hand the attempts of a burgeoning specialism to use the language of science and technology to bolster its claims and, on the other, its ability to situate this expertise within the social and political demands of an increasingly visual Victorian world. As illustrated in the exploration of both schools and the workplace, ophthalmologists undertook a distinct approach that involved their growing intervention at both a governmental and popular level. In their study of schools, they increasingly carried out governmental enquiries. In the workplace, they undertook an important advisory role for the Board of Trade. In both instances, ophthalmologists argued that the specificities of each environment demanded a certain level of vision. This was not calibrated to the environment itself and instead was formulated and created in response to the new 'normalised' standard. Employees on the railway were not expected to have a level of acuity in relation to, for example, distinguishing railway signals at a safe distance. Instead, they were expected to have 'normal' vision in one or both of their eyes. The same also applied to vision testing standards in schools. However, as this chapter has shown, 'normal' vision was not a fixed category; it was an arbitrary standard described as 'sufficient'. 'Normalcy' was not the physical norm, and the potential variability of the eye was interrogated continually as standardised measurements were popularised. Despite people's lack of conformity, or the existence of variability, this sufficiency determined and defined the parameters of appropriate vision. Testing was in its infancy in both schools and the workplace, but medical authorities emerged as experts, offering consultation and advice on vision measurement in line with new medical theory and practice. In doing this, they placed the problem on the individual who was expected to conform. The new standard had the potential to be misapplied but nevertheless increased the importance of technology in both identifying and also enhancing a person's visual capacity and removing the 'problem'. Medical practitioners conceived of a social norm, coined in an attempt to create a generalised, useful, standard that could sufficiently meet the needs of its time. Fixed and contestable in nature, this type of norm exposed the individuality of, and differentiation between, people's vision, which did not map neatly onto the 'normal' and 'abnormal' projection of vision across a person's lifetime. The corresponding rise in spectacle wearers explored in the following chapters only served to evidence this further.

Notes

1. Delia Gavrus, 'Men of dreams and men of action: neurologists, neurosurgeons, and the performance of professional identity, 1920–1950', *Bulletin of the History of Medicine*, 85 (2011), 57–92; Jaipreet Virdi-Dhesi, 'From the hands of quacks: aural surgery, deafness, and the making of a surgical specialty in 19th century London' (2014), unpublished PhD thesis.
2. Luke Davidson, '"Identities ascertained": British ophthalmology in the first half of the nineteenth century', *Social History of Medicine*, 9:3 (1996), 313–333.
3. Roger Cooter, *Surgery and Society in Peace and War: Orthopaedics and the Organisation of Modern Medicine, 1880–1948* (Basingstoke: Palgrave Macmillan, 1993), pp. 4–7.
4. George Weisz, *Divide and Conquer: A Comparative History of Medical Specialisation* (New York: Oxford University Press, 2006).
5. Lennard J. Davis, *Enforcing Normalcy: Disability, Deafness and the Body* (London: Verso, 1995).
6. Coreen McGuire, *Measuring Difference, Numbering Normal: Setting the Standards for Disability in the Interwar Period* (Manchester: Manchester University Press, 2020).
7. Peter Cryle and Elizabeth Stephens, *Normality: A Critical Genealogy* (Chicago, IL: University of Chicago Press, 2017), p. 3.
8. Ibid., p. 10.
9. Georges Canguilhem, *The Normal and the Pathological* (New York: Zone Books, 1991).
10. Ibid., p. 161.
11. F.C. Donders, *On the Anomalies of Accommodation and Refraction of the Eye, with a Preliminary Essay on Physiological Dioptrics*, trans. William Daniel Moore (London: New Sydenham Society, 1864).
12. See, for example, Mary Carpenter, *Health, Medicine and Society in Victorian England* (Santa Barbara, CA: Praeger, 2009), pp. 143–144; Wolfgang H. Vogel and Andreas Berke, *Brief History of Vision and Ocular Medicine* (Amsterdam: Wayenborgh Publishers, 2009), pp. 221–222; Daniel M. Albert, 'Ocular refraction and the development of spectacles', in Daniel M. Albert and Diane D. Edwards (eds), *The History of Ophthalmology* (Oxford: Blackwell Science, 1996), p. 75.
13. Mentioned in note to 1962 reprint of Donders, *On the Accommodation and Refraction of the Eye*, p. 278; in Wellcome Library.

14 Stanley Joel Reiser, *Medicine and the Reign of Technology* (Cambridge: Cambridge University Press, 1978), pp. 45, 55, 68.
15 Ibid., pp. 45–46.
16 Ibid., pp. 50–51.
17 See the work of C. Richard Keeler and, in particular, 'The ophthalmoscope in the lifetime of Hermann von Helmholtz', *Archives of Ophthalmology*, 120:2 (2002), 194–201 for an overview.
18 See, for example, Johann Friedrich Horner, *On Spectacles: Their History and Uses* (London: Balliere, Tindall & Cox, 1887), pp. 12–13. For another example, see Robert Brudenell Carter, *Eyesight: Good and Bad: A Treatise on the Exercise and Preservation of Vision*, 2nd edn (London: Macmillan, 1880), p. 76.
19 Donders, *On the Accommodation and Refraction of the Eye*, p. 33.
20 Ibid., pp. 31–32.
21 Ibid., p. 71.
22 Joel Soelberg Wells, *On Long, Short and Weak Sight and Their Treatment by the Scientific Use of Spectacles* (London: J.A. Churchill, 1862), p. 26.
23 See, for example, Robert Brudenell Carter, *Eyesight in Schools: A Paper Read Before the Association of Medical Officers of Schools on April 15th, 1885* (London: Harrison and Sons, 1885), pp. 5–6, 8.
24 See, for example, D.B. Roosa, *Defective Eyesight: The Principles of its Relief by Glasses* (London: Macmillan, 1899), pp. 44–45.
25 Peter John Brownlee, *The Commerce of Vision: Optical Culture and Perception in Antebellum America* (Philadelphia: University of Pennsylvania Press, 2019), pp. 106–109.
26 Donders, *On the Accommodation and Refraction of the Eye*, p. 41.
27 H. Snellen, *Test-Types for the Determination of the Acuteness of Vision* (1862), p. 4.
28 Reiser, *Medicine and the Reign of Technology*, pp. 91, 191.
29 Donders, *On the Accommodation and Refraction of the Eye*, p. 77.
30 Ian Hacking, 'Biopower and the avalanche of printed numbers', *Humanities in Society*, 5:3 and 4 (1982), 279–295.
31 Davis, *Enforcing Normalcy*, pp. 24–26.
32 McGuire, *Measuring Difference*.
33 Cryle and Stephens, *Normality*, p. 10.
34 McGuire, *Measuring Difference*, pp. 5–6; L. Daston and P. Gallison, *Objectivity* (New York: Zone Books, 2007), p. 34.
35 Jaiproot Virdi-Dhesi, 'Curtis' cephaloscope: deafness and the making of surgical authority in London, 1816–1845', *Bulletin of the History of Medicine*, 87:3 (2013), 347–377.

36 Gavrus, 'Men of dreams', pp. 70–71, 79–81, 90–91.
37 Cooter, *Surgery and Society*, pp. 4–7.
38 Parallels can be drawn with audiometry. See, for example, Virdi-Dhesi's 'Curtis's cephaloscope' and Jaipreet Virdi and Coreen McGuire, 'Phyllis M. Tookey Kerridge and the science of audiometric standardisation in Britain', *British Journal for the History of Science*, 51:1 (2018), 123–146.
39 Roosa, *Defective Eyesight*, p. 12.
40 Wells, *Short and Weak Sight*, p. 111.
41 Horner, *On Spectacles*, pp. 9–10.
42 Carter, *Eyesight Good and Bad*, p. 59.
43 Roosa, *Defective Eyesight*, p. 3. A good description of nineteenth-century test-types in use by the 1880s for near and distant vision can be found in A. Stanford Morton, *Refraction of the Eye: It's Diagnosis and the Correction of its Errors: with a Chapter on Keratoscopy* (London: H.K. Lewis, 1881), pp. 16–17.
44 Reiser, *Medicine and the Reign of Technology*, various.
45 'Guest editorial: hypermetropia or hyperopia?', *Ophthalmic & Physiological Optics*, 35 (2015), 2–7.
46 Donders, *On the Accommodation and Refraction of the Eye*, p. 124.
47 See, for example, Wells, *Short and Weak Sight*, pp. 99–100.
48 Claire L. Jones, *The Medical Trade Catalogue in Britain, 1870–1914* (London: Pickering & Chatto, 2013).
49 James Weiss & Son, *A Catalogue of Surgical Instruments, Apparatus, Appliances etc.* (1863); *Illustrated Catalogue and Price List of Surgical Instruments* (1889); *A Catalogue of Ophthalmic Instruments and Appliances* (1898).
50 See, for example, Mayer, Meltzer & Jackson, *A Catalogue of Surgical Instruments and Appliances* (1885); *Catalogue of Surgical Instruments and Appliances* (1890).
51 Down Bros, *A Catalogue of Surgical Instruments and Appliances* (1885; 1887; 1889; 1892; 1894; 1896; 1900); *A Catalogue of Surgical Instruments* (1897); *A Catalogue of Surgical Instruments and Appliances, also of Aseptic Hospital Furniture* (1901).
52 R. Keeler, 'Antique ophthalmic instruments and books: the Royal College Museum; Part I Instruments', *British Journal of Ophthalmology*, 86:6 (2002), 602–603 (p. 603).
53 Henry Juler, 'On the best methods of diagnosing and correcting the errors of refraction', *British Medical Journal* (27 December 1884), pp. 1274–1275.
54 See, for example, Morton, *Refraction of the Eye*, p. 27.

55 Davidson, '"Identities ascertained"', pp. 331–332.
56 William Bowman, 'An address delivered at the opening of the section of Ophthalmology', *British Medical Journal* (13 August 1881), pp. 277–279.
57 Jaipreet Virdi, 'Medicalising deafness in Victorian London: The Royal Ear Hospital, 1816–1900', in Iain Hutchison, Martin Atherton and Jaipreet Virdi (eds), *Disability and the Victorians: Attitudes, Interventions, Legacies* (Manchester: Manchester University Press, 2020), pp. 73–91.
58 Richard A. Meckel, *Classroom and Clinic: Urban Schools and the Protection and Promotion of Child Health, 1870–1930* (New Brunswick, NJ: Rutgers University Press, 2013), pp. 23–30; Chris Otter, *The Victorian Eye: A Political History of Light and Vision in Britain, 1800–1910* (Chicago, IL: Chicago University Press, 2008), pp. 43–45.
59 Arthur Newsholme, *School Hygiene: The Laws of Health in Relation to School Life* (London: Swan Sonnenschein, Lowrey, 1887), p. 109; Christopher Hamlin, 'Review of John M. Eyler, *Sir Arthur Newsholme and State Medicine, 1885–1935* (Cambridge: Cambridge University Press, 1997)', *Bulletin of the History of Medicine*, 72:3 (1998), 564–566.
60 Otter, *The Victorian Eye*, p. 44.
61 Meckel, *Classroom and Clinic*, pp. 25–26.
62 Reference to this is also made by Otter, *The Victorian Eye*, p. 43.
63 W.W. Ireland, 'Eyesight and the Education Act', *British Medical Journal* (26 January 1889), p. 213.
64 Priestley Smith, *Short Sight in Relation to Education* (Birmingham: The Midland Educational Company, 1880).
65 Simeon Snell, *Influences of School Life on Eyesight* (London: Wyman & Sons, 1884).
66 Carter, *Eyesight in Schools*.
67 J.D. Hirst, 'Vision testing in London: a rehearsal for the School Medical Service', *Journal of Education Administration & History*, 14:2 (1982), 23–29.
68 See, for example, Smith, *Short Sight in Relation to Education*, pp. 32–33; Snell, *Influences of School Life on Eyesight*, pp. 5–12.
69 Jamie L. Bronstein, *Caught in the Machinery: Workplace Accidents and Injured Workers in Nineteenth-Century Britain* (Stanford, CA: Stanford University Press, 2008), p. 3.
70 See, for example, 'Carter on *Eyesight Good and Bad*', *Saturday Review of Politics, Literature, Science and Art* (15 May 1880), pp. 637–638;

'Book review', *The Academy* (3 July 1880), p. 7; 'Bibliophile's kalendar', *Book-Lore* (November 1885), p. 182; 'Defective vision in school children', *Freeman's Journal* (17 July 1896); 'London, Friday July 17', *The Standard* (17 July 1896); 'Defective vision in elementary schools', *The Derby Mercury* (22 July 1896).
71 'Eyesight and school life', *The Athenaeum* (24 October 1896), pp. 567–568.
72 'Aberdeen Educational Institute – the eyesight of children', *Aberdeen Weekly Journal* (27 January 1890).
73 'Leicester School Board – the eyesight of school children', *Leicester Chronicle and the Leicestershire Mercury* (11 November 1899).
74 Richard Liebreich, *School Life in Its Influence on Sight and Figure: Two Lectures* (London: J. & A. Churchill, 1878).
75 Smith, *Short Sight in Relation to Education*, p. 13; Snell, *Influences of School Life on Eyesight*, p. 5; Edward G. Loring, *Is the Human Eye Changing its Form Under the Influence of Modern Education* (publisher not identified, 1878), p. 7; see also relevant comments in late nineteenth-century periodicals: 'Multiple news items', *The Morning Post* (18 September 1884), p. 4; 'The elements of school hygiene', *The Practical Teacher* (September 1900), p. 124.
76 Meckel, *Classroom and Clinic*, pp. 23–30.
77 'Shortsightedness among children in school', *The Graphic* (3 June 1882). For general comment on increase in short sight and criticism of print see, for example, 'Short sight in Germany', *The Morning Post* (18 August 1881); 'Education and eyesight', *The Morning Post* (29 August 1881); 'Shortsightedness in Germany' (*The Graphic*, 10 September 1881).
78 See, for example, 'Myope', *Bentley's Miscellany* (July 1867), p. 594; 'Bibliophile's kalendar', p. 182.
79 Newsholme, *School Hygiene*; 'The elements of school hygiene', *The Practical Teacher* (September 1900), p. 124.
80 Carter, *Eyesight in Schools*, p. 5.
81 Ibid., p. 4.
82 Ibid., pp. 4–5.
83 Otter, *The Victorian Eye*, p. 44.
84 Hirst, 'Vision testing in London', pp. 23–24.
85 'Short-sighted children', *The Standard* (20 December 1888).
86 'Our eyes and eyesight', *The Yorkshire Herald, and the York Herald* (21 January 1892).
87 Arnold Lawson, 'Abstract of a report on the vision of children attending London elementary schools', *British Medical Journal* (18 June 1898),

pp. 1614–1617; whereas, Ireland, 'Eyesight and the Education Act', p. 213, for example, claimed that steps had been taken.
88 Smith, *Influences of School Life on Eyesight*, p. 12; 'round-shouldered' is also mentioned in newspaper reports on a lecture that refers to short-sightedness and its relation to brain stress: 'Sir J. Crichton Browne on "brain stress"', *Birmingham Daily Post* (18 February 1890); 'Nature and science', *The Leeds Mercury* (22 February 1890).
89 Quoted in a review of 'Lectures on the recent progress of the theory of vision by Professor Helmholtz and *Eyesight Good and Bad* by Robert Brudenell Carter', *The Edinburgh Review* (October 1881), p. 539.
90 Smith, *Influences of School Life on Eyesight*, pp. 20–21; see also, Carter, *Eyesight in Schools*, p. 13; Loring, *Is the Human Eye Changing*, p. 22; Robert Farquharson, *School Hygiene and Diseases Incidental to School Life* (London: Smith, Elder, 1885), p. 363; see also examples in popular print: 'How we see, hear and speak', *The Leisure Hour* (January 1889), p. 64; 'The increase of short sight', *The Pall Mall Gazette* (16 July 1889). Loring also discussed the wider international consequences of the deterioration of eyesight in children: *Is the Human Eye Changing*, p. 2.
91 'Untitled', *The Morning Post* (16 July 1896).
92 Peter Conrad, *The Medicalisation of Society: On the Transformation of Human Conditions into Treatable Disorders* (Baltimore, MD: Johns Hopkins University Press, 2007), p. 3.
93 D. Love, 'The vision of school children', *British Medical Journal* (25 March 1899), p. 763.
94 For the subjectivity implied in the use of medical tools more broadly, see Reiser, *Medicine and the Reign of Technology*, p. 183.
95 See, for example, W. Kitchiner, *The Economy of the Eyes: Precepts for the Improvement and Preservation of Sight* (London: Hurst Robinson & Co., 1824) and a comparable discussion of this by Brownlee in *Commerce of Vision*, p. 11; and Jordanna Bailkin's assessment of the work of George Wilson in 'Colour problems: work, pathology and perception in modern Britain', *International Labour and Working-Class History*, 68 (2005), 93–111 (p. 95).
96 Bronstein, *Caught in the Machinery*, p. 3.
97 Ibid.
98 For more information on Curtis, see Virdi-Dhesi, 'Curtis' cephaloscope'.
99 'Critical notices', *The New Monthly Magazine and Literary Journal* (May 1833), p. 102.

100 'Employments which injure the eyesight, No. IV', *The Saturday Magazine* (9 June 1838), pp. 222–223; 'Employments which injure the eyesight, No. V', *The Saturday Magazine* (16 June 1838), p. 230.
101 *The North British Review* (November 1856), pp. 181–182.
102 'A plea for the eyes', *Chambers's Journal of Popular Literature, Science and Arts* (5 June 1858), pp. 357–358.
103 'International Oculist Meeting', *Glasgow Herald* (3 August 1872).
104 See, for example, *The North British Review* (November 1856), pp. 181–182.
105 See, for example, 'Carter on *Eyesight Good and Bad*', *Saturday Review of Politics, Literature, Science and Art* (15 May 1880), pp. 637–8; 'Helmholtz and Carter on eyesight', *The Edinburgh Review* (October 1881), p. 535.
106 'Chips', *North-Eastern Daily Gazette* (11 March 1891).
107 Loring, *Is the Human Eye Changing*, see particularly pp. 21–25.
108 J.T. Arlidge, *The Hygiene Diseases and Mortality of Occupations* (London: Perceval, 1892), pp. 199–200.
109 Ibid., p. 199.
110 From a lecture to the Royal Dublin Society on the overuse of the eyes, *Freeman's Journal and Daily Commercial Advertiser* (19 March 1887).
111 Ibid; Dr Arthur Benson was an original member of the Ophthalmological Society for the United Kingdom. For an overview of his career see 'The late Dr Arthur H. Benson', *British Medical Journal* (23 November 1912), p. 1502.
112 Steve Sturdy, 'The industrial body', in Roger Cooter and John Pickstone (eds), *Companion to Medicine in the Twentieth Century* (London: Routledge, 2003), pp. 218–221.
113 Sarah F. Rose, *No Right to be Idle: The Invention of Disability, 1840s–1930s* (Chapel Hill: The University of North Carolina Press, 2017), throughout.
114 For an example of this in the context of twentieth-century social and welfare reform see Joanna Bourke, *Dismembering the Male: Men's Bodies, Britain and the Great War* (London: Reaktion, 1999).
115 Anne Digby, *The Evolution of British General Practice, 1850–1948* (Oxford: Oxford University Press, 1999), p. 247.
116 Sturdy, 'The industrial body', p. 219.
117 Otter, *The Victorian Eye*, also considered vision testing in the army and train driving: p. 45.
118 See, for example, 'Restrictions on the French press', *Daily News* (9 February 1863); 'Foreign and colonial', *The Leeds Mercury*

(10 February 1863); 'General news', *The Dundee Courier & Argus* (23 March 1863); 'Untitled', *The Dundee Courier & Argus* (18 June 1877); 'Spectacles in the French army', *The Leeds Mercury* (11 July 1877); 'Spectacles and moustaches', *The Huddersfield Daily Chronicle* (17 December 1877); 'Untitled', *The Belfast News-Letter* (27 July 1893); 'The monocular forbidden', *The Essex County Standard West Suffolk Gazette, and Eastern Counties Advertiser* (24 February 1894); 'Untitled', *The Dundee Courier & Argus* (6 October 1894); 'Untitled', *The Yorkshire Herald, and the York Herald* (2 February 1899).

119 See, for example, 'Military and Naval Medical Services', *British Medical Journal* (12 November 1881), pp. 798–799; 'The public services: the Naval and Military Medical Services', *British Medical Journal* (2 September 1893), pp. 544–545.

120 T. Longmore, *The Optical Manual: or, Handbook of Instructions for the Guidance of Surgeons in Testing the Range and Quality of Vision of Recruits and Others Seeking Employment in the Military Services of Great Britain* (London: HMSO, 1885), pp. iii–iv.

121 Kenneth Macleod, 'Remarks on the physical requirements of the public services', *British Medical Journal* (11 May 1895), 1021–1025 (p. 1021).

122 See Chapter One, section headed 'The practice of spectacle dispensing and medical alternatives'.

123 E. Warlomont, 'On the use of optometers for the examination of soldiers and workmen employed on the railroad', *British Medical Journal* (5 March 1881), 333–336 (p. 336).

124 Bailkin, 'Colour problems', p. 94.

125 C.D. Roberts, *The Detection of Colour-Blindness & Imperfect Eyesight by the Methods of Dr Snellen, Dr Daae, and Prof Holmgren: with a table of coloured Berlin wools and sheet of test types* (London: David Bogue, 1881).

126 'Defective eyesight in railway servants and seamen: deputation to the President of the Board of Trade', *British Medical Journal* (9 February 1895), pp. 315–316; for information on Bickerton's position see Thomas H. Bickerton, *Colour Blindness and Defective Eyesight in Officers and Sailors of the Mercantile Marine: A Criticism of the Board of Trade Tests* (Edinburgh: James Thin, 1890).

127 See, for example, 'Defective eyesight in railway servants and seamen', pp. 315–316; 'Colour blindness and defective eyesight in the personnel of the Mercantile Marine', *British Medical Journal* (18 May 1895), pp. 1112–1113; 'The eyesight of seamen: colour blindness and defective sight in the Mercantile Marines', *British Medical Journal*

(30 January 1897), pp. 292–293; 'The eyesight of seamen: colour blindness and defective sight in the Mercantile Marines', *British Medical Journal* (27 February 1897), pp. 537–538.
128 David M. Turner and Daniel Blackie, *Disability in the Industrial Revolution: Physical Impairment in British Coalmining, 1780–1880* (Manchester: Manchester University Press, 2018), see especially pp. 186–189; Ben Curtis and Steven Thompson, '"This is the country of premature old men": ageing and aged miners in the south Wales coalfield, c.1880–1947', *Cultural and Social History*, 12:4 (2015), 587–606 (pp. 597–601).
129 'A discussion on the vision of railway servants', *British Medical Journal* (9 August 1891), pp. 466–470.
130 J.B. Lawford, 'The visual tests for railway servants and mariners', *British Medical Journal* (23 March 1895), p. 641.
131 Henry G. Terry, 'Railway servants' eyesight: two cases of "hard lines"', *British Medical Journal* (25 July 1896), p. 234.
132 Bailkin, 'Colour problems', p. 100.
133 W.M. Beaumont, 'Continental and British vision tests for railway servants: a comparison', *British Medical Journal* (16 October 1897), pp. 1085–1086.
134 'Railway signalling', *Daily News* (6 January 1882).
135 Beaumont, 'Continental and British vision tests', p. 1085.
136 S. Johnson Taylor, 'The vision of railway officials', *British Medical Journal* (26 March 1898), pp. 815–816.
137 'Defective eyesight in railway servants and seamen: deputation to the President of the Board of Trade', p. 316.
138 'The eyesight of seamen: colour blindness and defective sight in the Mercantile Marines', *British Medical Journal* (6 February 1897), pp. 343–344.

3

Challenging (ab)normalcy: expansion in manufacture, design and access, 1851–1904

Visual aid designs have evolved to be diverse. Frames can be thick, bold and striking or thinner and delicate, lenses can be a more conventional round or oval, or a variety of geometric shapes, and the material of the frame can vary from metal to plastic and come in a variety of colours. Amongst this compendium of twenty-first-century eyewear styles there are some essential and easily recognisable design features: a lens holder for each eye, a bridge for the nose and a pair of side-arms that reach to the ear to help secure the frame to a person's face. Such a frame would have not been that familiar to the early Victorian wearer. Studies of spectacle design have traced this 'modern' shape of spectacles – a frame with side-arm attachments resting behind the ear – to the middle of the nineteenth century.[1] However, it was far from refined and, in developing this style of frame, the Victorian spectacle wearer would have witnessed significant changes in spectacle design and access. Matthew William Dunscombe, a Victorian optician based in Bristol, amassed a collection of spectacles and eyeglasses to document these changes before it was later destroyed in a fire. At the Victorian Era Exhibition in 1897, Dunscombe juxtaposed the display of 'Wig Spectacles' from the 'reign of George III' alongside a pair of 'Gold Hook-side Bridge Spectacles' dated to 1893. Wig spectacles, as the name suggests, were secured to the face via side-arms attaching into a wig, whereas hook-side bridge spectacles had longer side-arms that curved at the end to mimic the shape of the ear and fix securely behind it. Dunscombe compared the two frames at the Exhibition to document what he perceived to be a significant 'difference in style of work' and the 'progress' that had been made in spectacle design.² The transformation of spectacle frame and lens design in the Victorian period, as Dunscombe sought to demonstrate, was profound. In addition to the changing design of side-arms, the types of metal used for frames, the bridge shape and the lens shape altered considerably. Dunscombe's second collection, which he amassed a few years before his death, was similarly used to illustrate the progress of nineteenth-century spectacle design.³ Dunscombe's second collection now resides in the Science Museum's Optics collection. The careful and calculated collecting practices of Dunscombe can perhaps be contrasted with the haphazard job-lots collecting style of Sir Henry Wellcome, whose collection of spectacles and eyeglasses resides alongside Dunscombe's in the Science Museum's Ophthalmology collection. Nevertheless, Wellcome, too, as part of his wider project to showcase how the history of medicine was built on evolutionary principles, was careful in his selection of objects and wished to display design progress.⁴

The cupboards and drawers of spectacles and eyeglasses collected by Wellcome's agents – in the basement of the Science Museum's off-site store in Blythe House, south London when I began researching them – offer more than a visual documentation of the way in which spectacle design altered. The some 1,000+ spectacle and eyeglass frames are a striking encounter with the multitude of Victorian spectacle wearers. This chapter draws heavily on material culture by analysing the specificities of design in the Science Museum's two collections. Handling and observing Victorian spectacles evoked a range of emotions and initiated several lines of enquiry. With the motivations and objectives of these collectors in mind, how typical was this collection of objects and to what extent did spectacle design evolve? What would it have been like to wear them? In comparison to the almost weightless acetate plastic I was wearing, the cumbersome older styles were heavy on the hand, while the thinner, lighter, metal styles were cold to the touch. By drawing upon business records, patent records, opticians' texts and advertisements to support the object findings, I explore the realities of spectacle wear in this chapter. I investigate how manufacture expanded and eventually produced a light, thin style of metal frame, the motivations behind these design changes and how the demand and growth of spectacle use challenged and engaged with the newly defined medical categories of normal and abnormal vision.

Alun Withey has identified the period between 1700 and 1850 as 'transformative' in the manufacture of eyewear. He emphasises key epochs as part of this narrative: metallurgical innovation in the manufacture of modish metals such as steel and silver and, from the 1720s, the idea that spectacles were not abstract contraptions to be looked through but instead were to be 'worn' and fixed to the face using side-arms.[5] Wearing spectacles, as Withey rightly argues, has several implications for the fashioning of the body and related social meanings and attitudes towards eyewear that will be discussed in Chapter Five. However, this chapter will consider the ways in which the wearing of spectacle frames generated a host of new design concerns, with Victorian manufacturing innovations focusing on the reduction of weight, the enhancement of comfort and durability. This chapter departs from Withey's analysis by extending its focus to lenses as well as frames, to demonstrate that the period between 1850 and 1901 was a different, but equally transformative, one. Advancements in the medical understanding of the refractive state of a person's eye and vision testing expanded the functionality of spectacles, based on a set of particular conditions. Spectacles needed to fit securely in order to accommodate more complex lens-types, new long-distance activities and longer, more extended, periods of wear. While Chapter One and studies of spectacles in a European context have shown that spectacles have been readily available since the fourteenth century, the way in which they were manufactured altered fundamentally in the second half of the nineteenth century, which saw a shift to a mass-manufactured, uniform frame.[6] Spectacle manufacture was mechanised and increasingly powered by steam, not hand, and this had significant implications for the style of frame and the numbers produced.

Claire L. Jones, in her recent work on prostheses and commodity cultures, has argued that prostheses were familiar by the Victorian period, and social and cultural categories of disability and/or impairment could now be standardised if different functional capacities of the body required normalising, as could the degree of normalisation.[7] Alun Withey similarly situates spectacles within a much broader context: a range of devices being used from the eighteenth century to help mould or shape the body to conform to normative or ideal standards.[8] Little is known, however, about the relationship between the design and availability of spectacles and their relationship to the growing normative attitudes surrounding visual capacity and vision standards. The universality of spectacle design and the mass manufacture of spectacle frames and lenses conflicted with medical practitioners' parameters for normal and abnormal visual capacity. Spectacle frames could be produced by piece-work *en masse* and did not need to be uniquely created to suit an individual's face. Building upon Chapter Two's discussion of medical practitioners' increasing favour towards spectacles in treatment practice, this chapter demonstrates that a spread in spectacle usage and the development of a more uniform frame acted more to challenge than to consolidate medical attitudes. The number of spectacle users made the 'abnormal' visually 'ordinary'. Enhanced medical knowledge about spectacles' utility created an expectation that everyone should have access to spectacles which, without any clear way of demarcating 'old sight' from refractive vision errors or abnormalities upon the face, exposed the paradox of a more common abnormality.

Designing a multifaceted device

Alun Withey has argued that alterations in the design of eighteenth-century spectacles influenced the way that the frames were used. In the Victorian period the relationship was more dynamic and I argue that the changing usage and function of spectacles influenced the way in which the frames and lenses were designed. Opticians and medical practitioners recommended different styles of frame based on the eyewear's intended purpose. As we have seen in Chapters One and Two, the traditional function of spectacles had been to aid a person's vision during close work; wearers often peered through a convex lens to make the words on a page, or an object close to hand, appear clearer. While concave lenses were the expected type for long-distance vision, spectacles were less frequently used for this purpose. Advancements in understanding the refractive condition of the eye, discussed in Chapter Two, exposed the need, use

Challenging (ab)normalcy 127

and utility of lenses for enhancing both a person's close and their distant vision. The frame that held convex and concave lenses came in a diverse range of styles and varied, often according to the lens' function. In 1855, for example, London optician Charles A. Long recommended a 'pantoscopic frame' for 'long sight', which was a frame that did not sit directly in front of the eye and instead angled downwards to suit a person's posture when reading. In contrast, 'short sight' required oval frames, as near to the eye as possible, to view one's surroundings.[9]

To explain why manufacture expanded and the motivations behind changing spectacle design, it is essential to first ascertain how spectacle design changed in response to alterations in spectacle use. At the start of the twentieth century American ophthalmologist R.J. Phillips, whose work was published by British trades publication *The Optician*, located the transatlantic impetus to change the design of spectacles in the material form and make-up of the frame. Phillips suggested that the 'clumsy' frames of bone, horn and shell from the eighteenth century were being replaced by the improved mechanical construction of 'light metal' later in the nineteenth century.[10] As 'clumsy' as late eighteenth- and early nineteenth-century frames might have been, they were diverse and increasingly well adapted for a range of different activities. In 1819, Liverpool-based optical instrument maker Egerton Smith advertised 'single jointed', 'double jointed' and 'hand' spectacles, quizzing glasses, 'gogglets' and a variety of different types of lens at the back of his text, *Hints to the Wearers of Spectacles*:

EGERTON SMITH & CO.

RESPECTFULLY SOLICIT THE ATTENTION OF THE PUBLIC TO THE FOLLOWING ARTICLES, Of the most approved Construction and Manufacture, which are CONSTANTLY ON SALE, AT THEIR OLD-ESTABLISHED SHOP, NO. 18, POOL LANE, Liverpool.

OPTICAL INSTRUMENTS.

Best double and single jointed gold Spectacles, with pebbles or glasses.

Do. Double jointed stout silver Spectacles, with ditto, round and oval eyes.

Do. Do. Do.   With ditto, and slip sprints.

Do. Do. Do.   With ditto, and swivel joints.

Do. Particularly light for walking.

Do. Single jointed silver Do. Round and oval eyes, do.

Do. Double jointed tortoise shell Do. With silver joints, round and oval eyes, do.

Do. Single jointed   Do.   Do.   Do.

Do. Tortoise shell Hand Spectacles, peculiarly convenient for occasional reading, do.

Spectacles for couched eyes.

Gogglets, or Shade Spectacles, for warm climates.

Best double jointed steel spectacles, round and oval eyes.

Do. Single jointed   Do.   Do.   Do.

Spectacle Cases, mounted in Nourse skin and silver swages.

Do. Nourse and dog skin, plain mounted.

Do. Tortoise shell and silver swages.

Do. Tortoise shell, plain mounted.

Do. Fish skin,   Do.

Do. Morocco, with snap springs.

Do. Do.   With straps &c.

Concave and Quizzing Glasses, mounted in gold and silver frames.

Do. Do.   In tortoiseshell and horn boxes.

Reading and Burning Glasses, in various mountings.

Watchmakers' and Multiplying Glasses.

Gogglers, with white or green glasses, to guard the eyes from dust or wind.

Best achromatic Operas, elegantly mounted.

Common   Do.   In various mountings.

Brazil Pebbles, Periscopic, Green and best plate Glasses, ground into any frames at a few minutes' notice.[11]

This variety can be broken down into fundamental categories of spectacle design in this period:

- those without side-arms that would have been held in the hand or perched on the end of the nose, often for close work (hand spectacles and quizzing glasses)
- those with side-arms but in a fixed and shorter style (single-jointed spectacles), and
- those with side-arms that had an additional attachment in a bid to fix the frame more easily and securely to the face (double-jointed spectacles).

The variety of designs included in Egerton's advertisement also reflect the variety of intended functions. The more secure double-jointed spectacles were 'particularly light for walking', the gogglets shaded the eyes in 'warm climates', hand spectacles offered convenience for quickly balancing a frame upon the nose for 'occasional reading' and quizzing glasses were a more decorative single-lens device with an ornate handle to hold and peer through, often at a social occasion.

Spectacles, therefore, were not an unsophisticated device by the beginning of the Victorian period, on the advent of improvements in medical understanding of the eye and vision testing, and came with a range of attachments and forms to suit their wearer's vision needs. Functionality influenced their design, and that functionality took into consideration the use of optical lenses to aid close and long-distance vision in private and public settings. Nevertheless, despite the use of side-arms in spectacle design from the 1720s, placing the frame behind the ears was not necessarily an obvious solution to the problem of securing the frame to the face of the wearer. For much of the early nineteenth century and early Victorian period, innovation in side-arm design focused on extending the arms to fix somewhere upon the face or in the hair. The 'double jointed' spectacle frame referred to in Egerton's advertisement was not a singular category and the term denoted several different styles of frame. The Science Museum's Ophthalmology and Optics collections include an array of 'transverse folding', 'extending', 'turn pin' and double-jointed frames. The name of the style of frame described the type of joint used; transverse folding frames (figure 3.1) had an

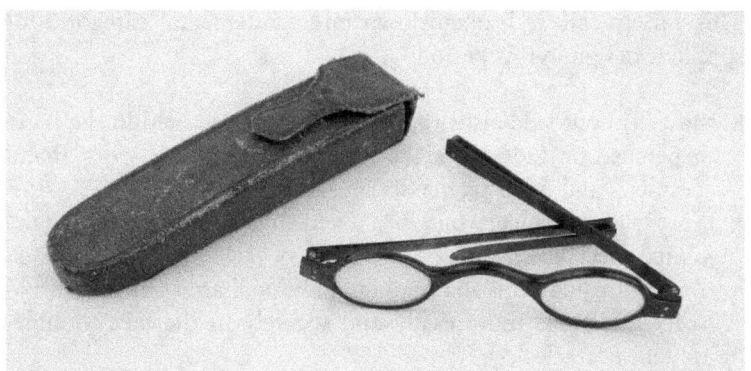

Figure 3.1 A pair of tortoiseshell transverse folding spectacles with the side-arm extensions closed. Dated to the early nineteenth century.

Source: Science Museum Art collection, 1948-397/44. With permission from SSPL/Science Museum.

extending attachment that folded out, extending frames (figure 3.2) had a sliding mechanism that lengthened the frame when in use and turn-pin frames (figure 3.3) had a pin that enabled an extension to the side-arm to be turned out.

Previous research has shown how these earlier frames were slowly displaced by 'straight' (figure 3.4) and 'hookside' (figure 3.5) side-arms.[12] Dunscombe's collection of spectacles and eyeglasses displays a similar pattern. Indeed, the use of the term 'double jointed' persisted in early Victorian advertisements before later being replaced with 'curved side' frames.[13] The older, heavier, extending and transverse-folding frames dated to *c.*1800 evolved into the thinner turn-pin frame, before being replaced with the steel 'wire', straight frame of the early Victorian period. By 1893, these metal wire frames were coil-sprung for comfort and elasticity, to fix more securely behind the ear.

Material evidence and opticians and medical practitioners' texts demonstrate that the heavier metal transverse and extending style frames quickly faded from use by the Victorians for functional reasons. Older spectacle frames in the Science Museum's collection abound with a variety of makeshift and more standardised techniques to improve the frame's comfort. Several thinner-style

Challenging (ab)normalcy

Figure 3.2 A pair of extending steel spectacles with a steel case and the right side-arm extended. Dated to the 1790s.

Source: Science Museum Ophthalmology collection, A682622. With permission from SSPL/Science Museum.

tortoiseshell straight spectacles dating from the late eighteenth century to the mid-nineteenth century, for example, have felt-like material attached to the side-arms or end-pieces of the side arm frames. Serrated marks on a further thirty-five tortoiseshell frames reveal the remnants of material that, through the passage of time or receipt of damage, has been worn or removed. In contrast to this more uniform design feature, a variety of tortoiseshell and metal

Figure 3.3 A pair of silver turn-pin spectacles with the right turn-pin side-arm mechanism slightly open. 'Improved pebbles' is visible on the left side-arm. Dated to the early nineteenth century.

Source: Science Museum Ophthalmology collection, A682430. With permission from SSPL/Science Museum.

frames provide evidence of individual attempts to adapt their frame. The spectacle bridge could be wrapped with padding in instances where there was no obvious sign of damage or as part of a user's attempt at a makeshift repair. The end-pieces of spectacle side-arms, too, are bound with soft material, which would have provided additional comfort to the user where the frame meets the face (at this time frames did not extend to fit and curl behind the ear). Such evidence leaves a physical trace of the pressure the frame exerted on its owner's face and their concerted efforts to alleviate this. It is perhaps not surprising, then, that in 1840 London optician John Thomas Hudson, for example, recommended the use of the 'turn pin' frame over the transverse folding and extending spectacles.[14] By 1880, ophthalmologist Robert Brudenell Carter only mentioned the turn pin and 'curled frame', another term for the hook-side frame; the heavier metal, transverse folding and extending types of spectacle were neither discussed nor extant.[15]

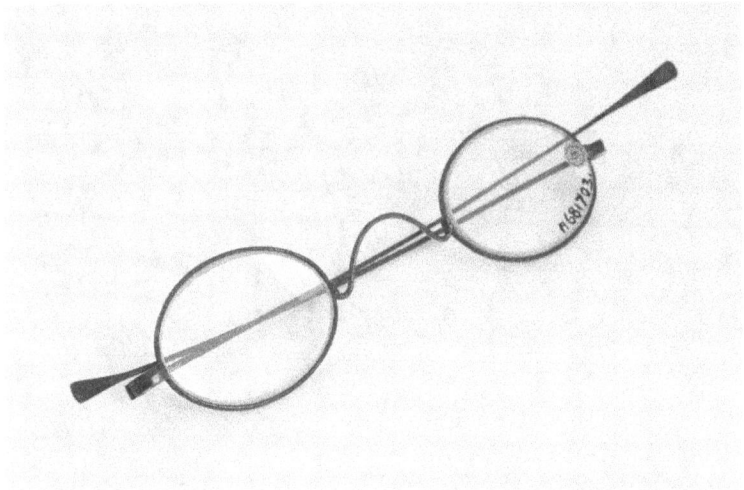

Figure 3.4 A pair of blued steel wire straight spectacles, typical of mid-nineteenth century styles.
Source: Science Museum Ophthalmology collection, A681703. With permission from SSPL/Science Museum.

The typical Victorian spectacle frame was thinner in style and increasingly featured a single side-arm piece that either was straight or sought to curve at the end to fix behind the ear. Analysing the average weight of 709 spectacle frames in the Science Museum's collections suggests that spectacles did get lighter, and probably more comfortable, across the century as a result of these design alterations. Older styles were between 10 and 20gm heavier than newer designs. The average weight of the transverse folding frame, for example, was 26.29gm, while the coil spring spectacles of the 1890s weighed 9.87gm. However, a reduction in frame weight would have been appreciable earlier than the 1890s. Comparing the average weight of the older steel and tortoiseshell spectacles, which would have been produced through cutting or stamping, with that of the steel wire straight frames which emerged from the 1830s reveals a 47 per cent reduction in weight. While such analysis is perhaps crude and cannot take into account the weight or strength of the lens for conservation reasons, it does propose that spectacles were becoming lighter.

Figure 3.5 A pair of steel-wire, coil-spring, hook-sided spectacles. Dated to 1893.
Source: Science Museum Optics collection, 1921-323/212. With permission from SSPL/Science Museum.

Lightness was a feature of spectacle frames sought by both opticians and medical practitioners. Yet the lightness of the frame was not simply an outcome of medical aspirations and evolved as a result of manufacturing innovations and the marketing of what was newly possible. Did medical ideas about vision influence this change or was it an outcome of broader changes? Who shaped spectacle design? The reality was a conglomeration of medical practitioners practising and exerting influence and opticians designing and advertising.

The persistent availability of a certain style of assistive aid can be a marker of their utility or of their reputation as effective or desired.[16] Manufacturing innovations made the light and elegant frame possible, but spectacle sellers and designers marketed and popularised it. Newspaper advertisements in a variety of locations pushed the 'peculiar lightness', or 'lightness and elegance of shape', in the early Victorian market between 1837 and 1864. To the early Victorian a lightweight frame was an undesirable, flimsy device but to the late Victorian it had become a delicate, tasteful one. A lecture on the development of optics in *The Optician* in 1898 concluded, for example, that:

[i]t is only in this century that flexible and twisted wire side pieces have been introduced, and the advance in the handling of metal generally rendered it possible to make spectacle frames light and elegant.[17]

The lightness of the frame was vital for spectacles as a device able to remedy a variety of refractive conditions for a longer duration. But were they any more comfortable with this longevity in mind? A surviving rubber end-piece on the end of a straight steel-wire spectacle arm (it is small, shaped like a boxing glove, and depicted in Figure 3.6) suggests that adaptions were made on and adornment added to lighter frames to improve their comfort.

In this period frame makers took out patents on a variety of materials to be added to spectacle frames to alleviate discomfort by the prospective spectacle wearer. Both new material technologies and more traditional materials were adopted. The catalogue for Dunscombe's first collection, for example, included a tortoiseshell

Figure 3.6 A pair of steel wire spectacles, typical of mid-nineteenth styles. The right side-arm has a surviving piece of material that would have been used to enhance the comfort of the end-piece.
Source: Science Museum Ophthalmology collection, A681780. With permission from SSPL/Science Museum.

plate on the bridge for 'tender skin', and the more modern celluloid on the side-arms 'to prevent contact of metal with the ear'.[18] India-rubber and rubber tubing appeared in a variety of patents between 1884 and 1900, to be attached to the bridge or side-arms of spectacles to prevent irritation.[19] Patent specifications emphasise the decisions behind the choice of material and design characteristics.[20] While metal spectacles were lighter, innovators stressed the potential damage to the face of wearing metal for lengthy periods of time. In 1892, *The Optician* featured what they described to be an 'ingenious application': A. Davison's newly secured 'cork-bridge' patent for spectacles. Davison argued that:

> [m]etal bridged spectacles... are a source of great pain and suffering to me, since by hurting and abraising the skin serious trouble has followed from their use, and [they] have been even known to have fatal results.[21]

Davison's statement about his own invention and the 'fatal' effects of metal bridges should be treated cautiously; but it does touch upon the potential risks of abrasion, rubbing and infection from broken skin on the nose arising from metal contact. It also evokes the potential discomfort associated with spectacle wear. Adaptions or adornments needed to be made in cases where people could not 'bear the weight of spectacles on the bridge of the nose'. Not all innovations were patented, however.[22] In addressing this potential 'wrinkle in spectacle fitting', London-based optician James Aitchison offered an unpatented solution: metal struts fitted to the lower rims of spectacles to rest on the greater surface area of the cheek bones.[23]

As 'ingenious' as Aitchison's cheek weight-bearing spectacles might have been, a fellow London-based optician argued that it would 'appear rather awkward on spectacles that have to be worn constantly'.[24] The design of spectacles was intimately bound up with function and, while opticians generated new designs, medical practitioners attempted to influence the style and type being produced. The heavily critiqued range of eyeglasses – a frame without side-arms – expanded substantially from 1850. The 'hand spectacles' in Egerton's advertisements had evolved from one of the oldest styles of spectacle, the rivet spectacle, able to perch and balance on the nose, primarily to assist in reading. The hand spectacles of the

Challenging (ab)normalcy

Victorian period needed to accommodate longer-term wear, as well as a broader range of active and static activities. The *pince-nez* style, which gripped the nose to stay in position, became increasingly popular from 1850. Indeed, although advances in eyeglass design occurred much later than advances in spectacle design, they were quicker. In 1855, 'hand folders' were riveted and had no spring.[25] Ten to fifteen years later these styles of frame incorporated a variety of different types of springs and plaquets (nose pieces) to accommodate different shaped faces and noses.[26]

The design of eyeglasses and debates concerning their use reveal the changing expectations of Victorian spectacle wear. Eyeglasses needed to be comfortable, secure and convenient. Patents in the 1880s and 1890s focused on achieving a delicate balance between a secure frame and the adjustment of the frame's spring pressure to prevent 'undue strain' or 'pinching'.[27] Advertisements, too, sought to focus on their frame's 'anti-pressure' qualities or their ability to remain 'firmly' upon the face 'without marking'.[28] In 1892, Messrs Wood (late Abraham) of Market Street in Manchester boasted that their 'Spectalette' could be 'used without intermission for a

Figure 3.7 A pair of blued steel *pince-nez* eyeglasses. Dated to 1860.
Source: Science Museum Ophthalmology collection, Y2000.246. With permission from SSPL/Science Museum.

whole day, and not produce any annoying or unpleasant effects'.[29] The comfort of the frame on a person's face is foregrounded in these discussions, but the positioning of the frame to ensure unpressurised contact with the skin was not the only consideration. A comfortable fit was important to ensure the angle in which a person looked through the lens was accurate. Improvements in understanding of the eye and lens manufacture – including the introduction of cylindrical and bifocal lenses – meant positioning the lens to a person's eye with greater precision became vital. Dunscombe, as part of his optical practice, proudly produced a pamphlet documenting the evolution of, and his contribution to, the improvement of the bifocal lens.[30] Previous work on the history of spectacle has tended to track the key developmental stages of these lenses in a similar style, often focusing on debates about their origin and invention.[31] Scholars, however, have not considered how the introduction of more complex lenses influenced frame design. The position of the frame determined the position of the lens, which, in the case of cylindrical and bifocal lenses, had different refractive powers in different focal planes and distinct fields of vision. Where a person looked through the lens therefore became increasingly more important for enhancing a person's vision.

Contemporary debates on the use of spectacles and eyeglasses show how important precision had become in the placement and wear of spectacles to ensure comfort, both on the skin and for the eyes. Advertisements for frames that do not 'shift from their position' or 'slip from the face' can be found from the 1850s.[32] Medical practitioners criticised the use of eyeglasses, because of the ease with which they could be displaced, or the inability to consistently ensure that the lens was in the centre of the pupils.[33] For brief usage, nevertheless, such portability also could serve a function. In 1880, for example, British ophthalmologist Robert Brudenell Carter described the convenience of a *pince-nez* as 'an accessory', but that it should not be 'the chief reliance of its owner'.[34] Opticians took medical practitioners' stance into account in their design considerations and convenience became an important design feature. Convenience, in scope to take eyeglasses on and off quickly and efficiently with one hand, was considered in both patents and the discussion of new inventions in *The Optician*.[35]

However, Carter advised the use of spectacles for permanent wear because they were 'correctly centred' and allowed the wearer to 'run, ride, dance or perform any other movements, without the glasses becoming displaced'.[36] While eyeglasses were convenient, spectacles offered the security required to adapt them to a range of physical activities and refractive eye conditions and this became more influential in eyeglass design considerations. In a bid to be more than just convenient, eyeglass design in the 1890s therefore focused on trying to increase the security of the frame and lens. Comments on new inventions emphasised their frame's newfound ability to 'grip perfectly' or reduce the 'tendency to shake' or 'rock'. Newly patented styles attempted to improve the frame's security by improving the stability of the bridge by using a more rigid joint or additional gripping device, or through mechanisms that enabled the bridge to be adjustable to the shape of an individual's nose. Implicated in the alterations to eyeglass design was the broadening function of frames and lenses. In 1894, a new visual aid design was discussed in *The Optician*. This came with 'detachable curlsides' so that it could function as both spectacles and eyeglasses. The 'curlsides' enabled the frame to be used on occasions that demanded a more secure fit, which encompassed 'horse-riding, playing cricket, tennis, music, cycling etc'.[37] Rather than providing only a means to continue close work, a visual aid – whether spectacles or eyeglasses – was expected to alleviate a variety of refractive eye conditions and accommodate a number of activities. While medical ideas certainly influenced the need for a sturdy frame, opticians were also responding to their own internal concerns and observations about the best available product for the prospective customer; the Victorian visual aid was designed in response to a much broader societal expectation for an adaptable, comfortable and affordable frame and lens.

A frame for the humbler classes: manufacturing expansion and access

The materiality of Victorian spectacle and eyeglass frames and alterations in manufacturing processes influenced their overall design and shape. Modish metals were introduced into the design and manufacture of spectacles in the eighteenth century and transformed the design possibilities of a variety of assistive devices.[38] However, it was not until mid-nineteenth century that the manufacture of spectacles began to substantially move away from tortoiseshell and horn frames.[39] In 1897, Matthew William Dunscombe wrote in the catalogue of his first collection that tortoiseshell was considered 'for 150 or 200 years... a favourite metal for the frames of the best spectacles, being soft and pleasant to the skin'.[40] Though comfortable on the skin and malleable, changes in manufacture made tortoiseshell less conducive to spectacle wear. By 1850, silver, horn and tortoiseshell were out of use, in favour of steel and gold, two metals that allowed spectacle and eyeglass frames to be mass-produced. The mechanisation of manufacture in the Victorian period created a new type of steel and gold frame that was both standardised and suited to large-scale production. By the 1890s, these metals were beginning to be replaced by alloys, advertised as aluminium or nickel, that further enhanced the functionality of the frame. The Science Museum's collections, for example, include a variety of plated, nickelled and gilt-steel frames designed to help prevent the frame from rusting.

Alterations in the manufacture of metal frames had arguably the most decisive influence on the form, availability and use of spectacle frames. When looking at the materials that have been catalogued in the Science Museum's collections, for example, steel has been separated into 'steel' and 'steel wire', two fundamentally different frames manufactured in different ways and that produced frames of a different thickness. As scholars have shown, developments in science and technology have had an influence on the material and form of prostheses and assistive devices.[41] Patents reveal that innovation in steel manufacture had a similar influence on spectacle frame design. While spectacles, like other assistive technologies, had taken advantage of cast steel's elasticity and malleability, the use of steel in spectacle manufacture did not reach its full potential until the Victorian period. Patents in the 1850s, for example, focused on a new technique of stamping and cutting out shapes from sheets of metal. Spectacle frame production quickly incorporated this process and, in 1854 and 1861, patents described variations of it utilising 'dies or cutting tools and a stamp or press'.[42] While such innovations created a thinner frame and more streamlined production process, improvements to the apparatus used to create thin wire spectacle frames enabled the manufacture of spectacles to reach an unprecedented scale. A patent from 1884, for example, described a process that would create spectacle frames out of a single piece of steel wire.[43] The production of steel-wire spectacle frames – which involved wrapping metal around a cylinder – could be mechanised on a larger scale. In 1895, a visitor to the factory of Newbold and Bulford described the manufacture of visual aids in the factory's workshops, reporting in *The Optician* that:

> [t]o get the frames to assume the round wire-like form so much in vogue, the strips, as we have described them, are run through a series of perforations in metal plates on a draw bench, which is essentially a combination of a vice and a pair of very formidable pincers, which seizes the rough strip of metal, and being attached to a strong leathern band winding round a roller, which is actuated by a lever, it draws the length of metal through the series of graduated perforations till the required gauge is reached.[44]

The mechanisation of spectacle frames in British factories was part of a rapid expansion in spectacle frame production, and style, in Europe and America.[45] Germany became a *locus* for spectacle manufacture and production, with several German optical factories – that would later become recognisable brands – being established from 1846, including Nitsche & Gunther, Rodenstock and Zeiss.[46] Scholars have estimated that in the mid-nineteenth century, 7.5 million spectacles were being produced per annum in the regions of Nuremberg and Furth.[47] While on a smaller scale, the manufacture of spectacles in Britain increasingly shifted to wholesale companies. From the mid-century, for example, Dunscombe's own Bristol-based optical firm supplied wholesale to Manchester, York and London.[48] Another optical firm in the Science Museum's collections, Sheffield-based Chadburn and Son, also manufactured on a large scale. Information about the firm's exhibit at the Great Exhibition of 1851 was displayed at the back of Alfred Chadburn's *Observations on the Choice and Use of Spectacles*. It positioned the firm at the cutting edge of spectacle and lens manufacture, arguing that its workshops produced 12,000 lenses a week by steam power.[49] The information they provided for the Children's Commission Committee offers an insight into their factory-based work and supports their technological exhibition display; the firm had a substantial workshop, machinery and overnight work.[50]

As Chadburn's exhibition hinted, manufacturers mechanised both lens and frame production. Like the choice of frame material, manufacturing processes altered the material recommended for lenses. Two types of spectacle lenses had been used interchangeably in the manufacture of early Victorian visual aids: glass, of various forms, and the more expensive Brazilian rock crystal, more commonly referred to as 'pebble'. The word 'Pebble' is engraved onto a variety of frames in the Science Museum's collections; the way it appears visibly on the side-arm of Figure 3.3 suggests that it was a marketable or desirable material. Indeed, spectacle sellers advertised 'best' pebble lenses at higher prices and emphasised their quality.[51] On a practical level, too, opticians and medical practitioners could recommend pebble because of its relative hardness and resistance to scratching.[52] However, pebble was also criticised because of its potential imperfections and its refractive index.[53] While there was no single consensus in opticians' and medical practitioners' texts over which lens material was the recommended choice, by the 1880s commentators tended to focus on the improved manufacture of glass, rather than the possible disadvantages of pebble. In 1880 Carter, for example, argued that glass manufacture had become more 'perfect', leaving little difference between glass and pebble.[54] The 'perfection' of glass manufacture in the Victorian period involved the refinement of mass manufacturing and bevelling machinery, with several patents focusing on both cylindrical and spherical lenses. In 1891, an article on 'Spectacle glasses' in a popular periodical recommended glass because of its improved manufacture:

> The perfection to which glass-making has now attained has rendered the use of artificial glass for spectacles practicable, and, seeing the advantages attaching to its employment, which it is the object of this short paper to point out, it is probable that it will entirely supersede rock-crystal.[55]

As we saw in Chapter One, the manufacture of lenses or frames by spectacle dispensers was increasingly more unusual in the

Victorian period. Traders therefore sought to entice customers by advertising their trade as a dying art. In 1847, a London optician proposed that he was 'the only optician that manufactures spectacles on the premises'.[56] Such a claim, while inherently false, was not blind to a growing trend amongst opticians to simply buy ready-made spectacle stock. *The Optician*'s editors claim in 1891 that 'we have no hesitation' in stating that 'not a single Optician manufactures his own glass' was not quite so far-fetched.[57] *The Optician*'s outlook on the deskilling of opticians was perhaps bleak and they worried that the manufacture and repair of frames in larger workshops and factories by wholesale firms reduced the 'handicraft work' required by their trade. The reality, however, was a more blended process in which machinery and hand-work operated in tandem, particularly for more high-end styles. An employee of the London-based Aitchison Optical Works in the 1890s, for example, argued that in the last decade of the century Aitchison's company was run on the principle that 'the personal equation should, as far as possible, be uppermost' and that a model in which 'the machinery [was] subsidiary to the man had been carried out'.[58] In 1893 a description of Chadburn and Son's workshops similarly concentrated on the intertwining of 'a large number of highly skilled and experienced hands' working in extensive workshops that housed 'elaborate and specially constructive machinery'.[59]

In practice, Charles Booth's notes for the London optical firm Curry & Paxton indicate the number of people still employed in spectacle-frame manufacture, and the type of work involved in the 1890s. The questionnaire recorded a total of 129 employees employed in a range of different work, including that of a spectacle maker, folder maker, gold spectacle maker, gold folder maker, optical framer, glass grinder, glazier, engineer, carpenter and pointer.[60] The skill of its workmen is impressed on the readers of *The Optician* in 'A review of Curry and Paxton', which concluded that 'the large number of skilled workmen employed by this firm ensures the utmost promptitude as well as accuracy in the execution of orders'.[61] As the breakdown of employees for Curry & Paxton suggests, the organisation of labour in the manufacture of spectacles was similar to the scientific instrument trade at this time, which had become increasingly sub-divided and focused on piece-work.[62]

A breakdown of the employees for London optical firm C.W. Dixey in Charles Booth's notebooks substantiates this: five employees were paid by 'piece work' as part of their average earnings of 60s. per week.[63]

The way in which frames and lenses were manufactured serves to reinforce the unique characteristics of spectacles and eyeglasses as an assistive device and the differences between the creation of an item of luxury or functional goods. The rapid expansion in the manufacture and use of the artificial limb in the First World War, for example, did not result in a standardised product. As demand for artificial limbs grew because of war injuries, a type of standardisation was attempted, and not achieved, in the later stages of the war. The methods, mechanisms and materials could be standardised, but the limb itself was individual to its user. The case of the artificial limb reveals the role of demand in initiating standardising innovations but the concept of an 'identical object' was alarming to contemporary designers, manufacturers and users.[64] In contrast, the component parts of spectacles, and the frame in its entirety, could be standardised. A debate on ready-made spectacle frames in *The Optician* in 1898 instigated by the owner of C.W. Dixey, W.A. Dixey, reveals that alarm over ready-made frames was a minority view. Bespoke items manufactured through piece-work were atypical. Dixey had responded to an article on 'Frames', which advised readers on how to select spectacles for a person from a bought-in range of stock and adjust them to the individual's face.[65] He argued that a frame should be made to fit the person and not later adapted. Dixey's stance sparked several responses that sought to emphasise the expense, technicality and number of workmen required to produce custom-made frames, which is evident in his own employee records.[66] A 'student of optical handicraft' challenged Dixey; the student questioned why an appropriately adjusted, ready-made frame should be deemed 'unsatisfactory'.[67] Rather than a superior alternative, a large proportion of opticians perceived bespoke frames to be unnecessary and impractical. With the conclusion by some that 'a specially made frame is needed but rarely', opticians instead emphasised standardisation and the easy availability of 'interchangeable lenses', which necessitated uniform lens and frame styles.[68] Rather than alarm being directed to the introduction of standardised products, contemporaries concerned themselves with the affordability of the frame and practicality of workmanship. Bespoke items were not the norm; instead, the journal documented accounts of unusual-shaped faces or customers that would require, in special circumstances, frames to be specially made or sent back to the spectacle maker.[69] Such a stance reveals that the expansion in spectacle use was not just about supply and demand but the characteristics of the frame and lens as an assistive device. The components of the frame – thin metal wire – were more conducive to adaptation to the individual face following methods of mass production in such a way that met the satisfaction of spectacle dispensers.

Whether spectacle lenses were ever truly 'interchangeable' is open to question. But calls for standardisation and uniformity reflect both the manufacturing process and an acute awareness of an increasingly divided spectacle market. The employee records of C.W. Dixey and wholesale firms such as Curry & Paxton reveal the number of workmen required to produce spectacle and eyeglass frames, which was a key factor behind the manufacture and retail of spectacles becoming increasingly more distinct. C.W. Dixey's, Aitchison's and Chadburn and Son's frames were cost-intensive and highly specialised, bespoke, luxurious items that had a narrow market. A receipt shows a pair of 'best quality curl side nickelled spectacles' with 'pebble lenses', for example, were bought from Aitchison for £1 5s. in 1899.[70] This did not cater for a growing number of spectacle users that were seeking good-quality, but affordable, frames and lenses. Was the ready-made frame more affordable? In 1877 Friedrich Horner, Swiss ophthalmologist and professor at the University of Zurich, argued in his reflections on the global history of spectacles that manufacturing developments directly correlated with a reduction in their cost:

> Now-a-days the processes of the manufacture of spectacles have been so greatly improved that they can be sold at a price that all can afford.[71]

Since the mid-eighteenth century there had been a variation in the materials, styles and quality of visual aid design, which meant spectacles could be retailed at different prices. Early nineteenth-century catalogues of manufacturing and wholesale opticians reveal that price was affected by the material of the frame and lens and relative demand, with manufacturing costs being reduced when frames were purchased by the dozen.[72] However, a pair of spectacles still cost over a shilling and there were early calls for charitable intervention to broaden accessibility. In 1835, an excerpt from one of the founders of ophthalmology in Britain, William Mackenzie, with an early but cautious support of spectacle use, appeared in a London periodical with the argument that 'there could not be a more useful appropriation of the funds of charitable institutions than that of providing spectacles for the poor'.[73] Both Mackenzie's earlier and Horner's later observations indicate an implicit awareness that medical ideas and their influence on spectacle use depended on the increasing affordability of spectacle and eyeglass frames. By the late 1870s and early 1880s, Horner was supposing that alterations in the scale of spectacles made spectacle wear accessible for everyone. A catalogue for Chadburn and Son in the 1890s perhaps substantiates this. It lists fine gold spectacles as well as a cheaper steel spectacle frame which could be purchased for only a shilling, a price which did not exist in the early Victorian catalogue. The collecting styles of Dunscombe and Wellcome embody the increasingly divaricate nature of the Victorian spectacle market. Dunscombe's collection is a careful, calculated one focusing on 'choice' or high-end items that display finer workmanship.[74] In contrast, Wellcome's acquisition records reveal that many spectacles were bought as part of cheap job-lots at auctions, which included a range of everyday objects collated under terms such as 'sundries'.[75] This sundry form of sale enabled spectacles to be sold, not for shillings, but for pennies.

Scholars have paid little attention to the cost of Victorian spectacles but have predicted that the cost of spectacles and eyeglasses reduced during the Victorian period. Studies have often assessed cost speculatively and have drawn upon singular examples in advertising material without factoring in inflation.[76] Real price change – which considers a consumer price index to ensure prices are comparable across time, as opposed to a nominal price change that simply compared prices as quoted – can lead to substantial differences in findings. Comparing the cost of spectacles sold in the cashbooks of two opticians and jewellers operating for the periods

1837–1851 and 1885–1897, for example, reveals a nominal price change of −18d. in comparison to a real price change of −12.2d.[77] It is also difficult in singular examples to assess the influence of location and/or position of the seller in the retail market. In this comparison, both businesses were comparable suppliers selling a similar quality of spectacles, but what about location? John Potter Dowell retailed in Carlisle and Robert Sadd retailed in Cambridge. Would this have a bearing on price comparison and were Carlisle and Cambridge cheaper or more expensive than each other (or London)?

Keyword searching the digitised popular press has made it possible to utilise a much larger sample size to address some of these questions of location, price and variation in market range. The new sample, while still limited in its scope, would previously have been unobtainable. An earlier, comparative study by Jonathan S. Pointer, for example, has shown how the price of spectacles remained akin to the daily average wage over a 700-year period, between 1276 and 1996.[78] However, before digitisation, the amount of data that could be drawn upon between 1830 and the early twentieth century would have been heavily limited, and this would have affected the breadth of such a study. Analysing seventy-nine advertisements between 1829 and 1900 revealed that, although the mean price did not fall, spectacles were advertised at a lower price from 1830 and a low-end price below 1s. 8d. was consistently available across the century. Such a narrative of increasing affordability during the Victorian period should itself be treated cautiously. Between 1830 and the early twentieth century spectacle prices fluctuated. The low-end market for spectacles, for example, remained consistent in London but grew slightly in towns and cities outside the metropolis. In contrast, prices at the upper-end of the market of spectacles fell in both London and provincial towns before 1880 (as advertised in their newspapers) but, after 1880, provincial prices began to rise again while London prices continued to fall. Such findings reveal a greater discrepancy of price in the provincial market for spectacles.

While the cost of spectacles did not continue to fall across the Victorian period in an obvious manner, the availability of cheaper frames created a noticeable difference in the language and styling of retailers' advertisements. The consistently available, cheaper styles created a new market of spectacle users that retailers were keen to attract. Optician and jeweller Robert Sadd, for example, recorded in his cashbook a sale of 'common spectacles' in 1837. Sadd also recorded the sale of 'second hand specs' alongside the sale of 'fine' or 'best' frames in his ledgers.[79] While retailers had often included a cheaper style of frame in their advertisements since the eighteenth century, the specific targeting of the lower classes was a Victorian phenomenon. Such a phenomenon was different from advertisers' claims that their products were affordable because their prices were 'very reduced', 'moderate', 'unusually low' or 'the lowest charged', which continued to abound in Victorian spectacle sale.[80] Retailers specifically identified social groups for the sale of certain products. In 1868, London-based Abraham & Co, for example, included an adjunct at the bottom of its advertisement: '***Spectacles for the Humbler Classes', while R.W. Mason of Leicester argued that its spectacles were 'suitable to all CLASSES' and that it had 'Special Terms for Working Classes'.[81]

Advertisers tended to make these cheaper items appear desirable by stressing their utility. 'The Working man's spectacle' was tagged to the bottom of an advertisement in *The Wrexham Advertiser* in 1887 while advertisers in Belfast and Glasgow included a range of 'good serviceable' and a 'thoroughly good' pairs of spectacles.[82] Scholars have identified the influence of class on prosthetic design, which often focuses only on functionality at the lower end of the market.[83] While the sale of spectacles suggests that visual aids followed this trend, there was a range in quality and concerns about the extent to which spectacles sold at a lower price were indeed functional. Abraham & Co, for example, had marketed their humble spectacles for between 1s. 6d. and 2s. 6d. and a 'thoroughly good' pair of spectacles could be bought in Belfast for 1s. In contrast, advertisers in Bristol and Dundee advertised cheap spectacle frames for 6d.[84] Reports of fraud in popular literature tended to focus on the 'rascality' of fraudsters preying on the 'working man' or 'poorest person' across the Victorian period.[85] Implicit to this type of discourse was the idea that spectacles were a 'basic right'. Similar to broader trends in the marketing of Victorian prostheses, the commodification of spectacles shaped attitudes to visual impairment and played a role in standardising how vision was, and should be, 'corrected'.[86] But the intended demographic of recipients for this form of visual aid charity broadened. Spectacles demanded a universal, cross-generational market. In the 1850s and 1860s, for example, the provision of spectacles for the elderly was considered important in the reports of a poor law inspector for the Southwell Workhouse in Nottinghamshire.[87] In contrast, medical intervention and the introduction of vision testing in schools, as discussed in Chapter Two, exposed the problem of affordable spectacles for children. The debate on who should provide vision testing, and how to tackle the growing numbers of children requiring 'assistance', in correspondence in *The Standard* in December 1888, had to contend with the fact that 'spectacles are beyond the reach of many parents'. As a solution, the correspondent predicted that there would be 'little difficulty' in providing children with them through charity provision.[88]

Additional costs also impinged on spectacle wear and the affordability of spectacles cannot be assessed purely through the one-time purchase of a single frame. Other expenses included the management and upkeep of a frame or lens, or the replacement of frames and lenses when either had broken or a person's eyesight had changed. A range of frames in the Science Museum's collections have evidence of damage and repair on the lenses, plaquets, side-arms, hinges or bridges. Aspects of the frame and lens have been fixed or reinforced using a variety of haphazard and more sophisticated methods. The physical remains reveal former users' choices and the delicacy of Victorian frames and lenses. Wrapping material around a broken spectacle bridge to make the frame rewearable, for example, highlights, on one hand, the number of bridges broken by spectacle users and, on the other, those users' choice to attempt a makeshift repair either because of a personal attachment to the frame or to avoid the cost of a more conventional repair. Opticians' ledgers indicate that they had a considerable repair caseload, often referred to generically as a 'spec job', or broken down into specific repairs to the side, frame, lens, plaquet, spring or screw.

As Table 3.1 shows, the individual cost to repair or purchase a new piece was relatively small with, on average, costs of repairs varying between 37d. for a new frame and as little as 3d. for a new

Table 3.1 Average cost of different types of repair from John Potter Dowell and Robert Sadd's account books.

Type of repair	Robert Sadd, 1837–1851 (price in pence)	John Potter Dowell, 1885–1898 (price in pence)
New lens	18.6	14.5
Repair spectacles	16.7	5.8
New sides to spectacles	28.2	10.1
Rep. eyeglasses	12.0	9.8
New spring	6.0	14.3
New screw	6.0	3.6
New frame	37.0	17.0
New bridge	22.5	10.5
Glasses repair	12.0	17.0
Gold repair	23.5	6.2
Folder repair		5.5
New plaquets		9.0
New piece		14.0
Reader repair	12.0	

screw. Cumulatively, however, the costs associated with spectacle wear could be substantial. The transactions of Reverend A. Wrigley between 1886 and 1892 amounted to £1 7s. for the purchase of eyeglasses, folders and spectacles, as well as folder repair, a new spring for his eyeglasses and a new curl and side for his spectacles.[89] Several customers made multiple visits for the purchase and repair of their spectacles and eyeglasses. While their backgrounds were not always recorded, they did include a range of professions such as draper and ironmonger.[90] The transactions reflect both the frequent need for repair and the need for different types of frame for different activities. Mr Rooper of Trim Street in Cambridge, for example, spent £2 10s. 6d. between 1868 and 1871 on the purchase of 'powerful convex' spectacles, tinted spectacles, reading glasses, cases and the repair of the frames.[91]

The repeated purchase of 'good serviceable' spectacles and repairs would have been considerably cheaper than the prices of optician and jeweller Robert Sadd (as revealed by the sales figures in his accounts), and the poor would have been unlikely to own different types of frame for multiple activities at any one time. Nevertheless, Sadd's accounts indicate the regular interaction between retailers and spectacle users and that spectacles frequently required changing. With this in mind, even cheap frames for pennies could become unaffordable. A description of a visit to the Royal London Ophthalmic Hospital – more commonly known as Moorfields Eye Hospital – in *The Leisure Hour* in 1890 revealed a charitable system based on need, in which some were provided spectacles 'free of cost' while to others, that could afford something, spectacles were 'sold at a very low price'. Obtaining spectacles, particularly in a hospital setting, had additional associated costs including treatment fees and travel to a more specialised and non-local site that patients, or parents of patients, could not always afford to do.[92] The number of those unable to afford spectacles, or associated costs as part of their treatment plan, is evident in a complaint to the Royal London Ophthalmic Hospital in 1900. The complaint argued that the spectacles and artificial eyes provided, which cost the institution £135 3s. 6d. per annum, had been improperly sourced from profiteering opticians and this negatively impacted the financial stability of the institution and its ability to become a self-supporting charity.[93] Such a standpoint, in addition to questioning the financial viability of free or subsidised spectacles for a hospital institution in a complex retail marketplace, serves to demonstrate the relative cost of, and demand for, good-quality, but affordable, spectacles.

The mass manufacture of frames, promotion of cheap spectacles and availability of medical charity fed into a growing medical and popular concern that the number of spectacle users was growing. In 1880, for example, *The Morning Post* reviewed a publication by Carter in relation to the 'prevalence of defective or impaired vision in the present day'.[94] Other newspapers drew upon medical commentators, public talks or government departments to explain a growing concern. In 1883, for example, *The Leeds Mercury* reported on the Health and Education Department which held the view that 'our nation is becoming prone, so much so that every ophthalmologist is constantly treating cases of myopia'.[95] In 1884 *The Morning Post* published a contemporary fear that future generations could become blind and that:

[s]pectacles, *pince-nez* and eye-glass, formerly relegated to the old or middle-aged, are now the ordinary and necessary appendages to be placed on the faces of young people of both sexes.[96]

The way commentators mapped the expansion in spectacle use onto contemporary understandings of normal and abnormal vision was paradoxical. The author in *The Morning Post* argued that a change in human activity was required to improve the 'abnormal shape' of the eye that so many people possessed, to restore vision. Since few statistical data survive, it is not possible to test one person's observations, but it is possible to observe that the 'abnormal' was being perceived as common. Spectacle use and mass manufacture, in addition to the adoption of the terms abnormal and normal vision in workplace and school settings, exposed the regulating standards as the ideal state, rather than the nominal average, of people's eyes.

Conclusion

In 1894 Alfred Chadburn concluded that 'amongst the improvements which time, experience and ingenuity have introduced into ordinary manufacture, few are more striking than those which affect the spectacle frame maker'. A visitor to the factory of George Culver in 1894, which produced the 'revluc' eyeglass *en masse*, argued that Culver's factory would:

> [m]ake an Optician who quitted the world a hundred, fifty, or even twenty-five years ago, stare and rub his eyes could we revivify him and transplant him suddenly to an optical factory carried on in the most improved manner of to-day.[97]

Such expansion was as a result of the particularities of the visual aid as an assistive device: the size, shape and material of the frame enabled it to be generically made and later adapted to a person's face to suit a range of conditions. Contemporaries presented the scale of spectacle and eyeglass frame manufacture as a positive narrative of achievement. In 1893 Chadburn and Son's factory, for example, had featured as part of a publication that documented a 'Century's Progress' of commerce in Sheffield, Rotherham and Mexborough.[98] Spectacle design was lighter, and spectacle manufacture was powered by technological innovations on a large scale. This forward march of achievement was offset, however, by what the mass manufacture of spectacles exposed. Several new technologies and environments – often similarly touted by Victorian contemporaries as markers of progress – affected people's vision and created a host of new prospective spectacle users. While medical practitioners with their new knowledge of the refractive condition of the human eye were well placed to receive a large population of new patients, the reality of number and type of prospective users affected the way in which 'normal' and 'abnormal' vision was culturally constructed and, as Chapter Five will show, the framing of refractive eye conditions as a 'disease', 'problem' or 'impairment'.

The numbers of spectacles being sold further complicated medical intervention because it created a problem from the viability and practicality of medical control over spectacle sale. As Chapter Four will explore, the sale and price of spectacles was not influenced just by the manufacture of frames and lenses in newly mechanised and cost-effective ways. Spectacles needed to be appropriately fitted and adapted to a person's face, a service that came with a cost and was subject to scrutiny. The London Spectacle Mission founded by Dr Waring in 1886 sought to supply spectacles 'to those of the poorer classes who stand in need of them' when their eyesight started to diminish in old age. In early newspaper clippings it was stated that more than 500 people were 'relieved' each year who would otherwise have been 'deprived of their means of livelihood by failing eyesight'. The charity was carried on by Dr Waring's daughter following his death. A response to a letter of enquiry in 1896 indicated that Miss Waring judged the kind of spectacles required except in cases where the patient had previously attended a hospital for their eyes where, instead, they would be given an order to take to the optician.[99] The recipient concluded that it would not 'be well' to use the service because Miss Waring appears 'to be the sole judge of the suitability of the spectacles'.[100] Later correspondence to the charity in 1899 and 1900 similarly asked whether the enterprise was 'genuine', of 'any use' or 'medically satisfactory'.[101] These concerns were not with the quality of products, which were 'properly tested' to ensure they were 'new, strong and well-made'.[102] They were directly concerned with how they were dispensed. The Mission was a sizeable charitable enterprise which continued to expand; newspaper clippings and the Mission's own pamphlets claimed that considerable numbers of poor people had been supplied with spectacles, including 1,516 in the year 1902 and a total of 37,000 between 1886 and 1912.[103] How did medical ideas affect sale? How did medical practitioners respond to, and attempt to control, the distribution of frames and lenses; or did they? The charitable provision of spectacles by the London Spectacle Mission included the provision of spectacles, a spectacle case, a Testament and a book of prayers by Dr Waring; but not the latest vision testing techniques. A cheaper frame was not always sought for in a hospital and the dispensing of spectacles was influenced by its well-established retail market and popular responses to vision loss that, as shown in Chapter One, did not see the wearing of spectacles as a medical problem.

Notes

1 H.W. Holtman, 'A short history of spectacles', in W. Poulet, *Atlas on the History of Spectacles*, trans. Frederick C. Blodi (Godesberg: Wayenborgh, 1978), p. xviii; B. Michael Andressen, *Spectacles: From Utility Article to Cult Object* (Stuttgart: Arnoldsche, 1998), p. 20.
2 Science Museum Technical File T/1921-323, Catalogue of the 'Exhibit of Spectacles at the Victorian Era Exhibition', 1897, p. 4.
3 Science Museum Technical File T/1921-323; Margaret Mitchell, 'Optics and the Science Museum', *The Optician* (21 September 1979), pp. 22–24 (p. 23).
4 See, for example, Ghislaine M. Skinner, 'Sir Henry Wellcome's Museum for the Science of History', *Medical History*, 30 (1986), 383–418; Ken Arnold and Danielle Olsen (eds), *Medicine Man: The Forgotten Museum of Henry Wellcome* (London: The British Museum Press, 2011), throughout. Wellcome was not alone in this evolutionary thought process at the time; see for example, Arthur MacGregor, 'Exhibiting evolutionism: Darwinism and pseudo-Darwinism in museum practice after 1859', *Journal of the History of Collections*, 21:1 (2009), 77–94.
5 Alun Withey, *Technology, Self-Fashioning and Politeness in Eighteenth Century Britain: Refined Bodies* (Basingstoke: Palgrave Macmillan, 2016), p. 92.

6 Vincent Ilardi, *Renaissance Vision from Spectacles to Telescopes* (Philadelphia, PA: American Philosophical Society, 2007), pp. 75–79, 128.
7 Claire L. Jones (ed.), *Rethinking Modern Prostheses in Anglo-American Commodity Cultures 1820–1939* (Manchester: Manchester University Press, 2017), p. 4.
8 Withey, *Self-Fashioning and Politeness*, pp. 1–17.
9 Charles A. Long, *Spectacles: When to Wear and How to Use Them: Addressed to Those who Value Their Sight*, 2nd edn (London: Bland and Long, 1855), pp. 21–23.
10 R.J. Phillips, *Spectacles and Eyeglasses: Their Forms, Mounting and Proper Adjustments*, 2nd edn (*The Optician and Photographic Trades Review*, 1900), p. 24.
11 Egerton Smith, *Hints to the Wearers of Spectacles; or a Concise Statement of the Comparative Merits of Pebbles and Glasses, When Used as Spectacle Eyes* (Liverpool, 1819), pp. 8–9.
12 Derek C. Davidson, *Spectacles, Lorgnettes and Monocles* (Princes Risborough: Shire, 2002), pp. 16–17; Hugh Orr, *Illustrated History of Early Antique Spectacles* (London: author, 1985), p. 38; William Rosenthal, *Spectacles and Other Vision Aids: A History and Guide to Collecting* (San Francisco: Norman, 1996), p. 111.
13 For 'double joints' see, for example, the *Examiner* (27 January 1839), p. 62; and for mention of 'single or curved sides', later in the century, see, for example, *Daily News* (10 September 1877); *The Wrexham Advertiser, and North Wales News* (26 March 1887).
14 J.T. Hudson, *Useful Remarks upon Spectacles, Lenses, and Opera-Glasses; with Hints to Spectacle Wearers and others; being an epitome of practical and useful knowledge upon this popular and important subject* (London: Joseph Thomas, 1840), pp. 12–13.
15 Robert Brudenell Carter, *Eyesight: Good and Bad: A Treatise on the Exercise and Preservation of Vision*, 2nd edn (London: Macmillan, 1880), pp. 244–246.
16 Graeme Gooday and Karen Sayer, 'Purchase, use and adaptation: interpreting "patented" aids to the deaf in Victorian Britain' in Jones (ed.), *Rethinking Modern Prostheses*, p. 40.
17 George Lindsay Johnson, 'The development of optics in the present century', *The Optician* (21 April 1898), p. 201.
18 Science Museum, 'Exhibit of Spectacles', p. 9. An example of a frame with a 'shell-lined bridge' is also present in the current Dunscombe collection at the Science Museum: object number 1921-323/210.

19 See for example, A.J. Boult, patent number 11,710 (27 August 1884); A. Davidson, patent number 15,928 (6 September 1892); J. Rintoul, patent number 6,127 (24 March 1894); D. Rugg, patent number 21,755 (13 August 1900).
20 Jaipreet Virdi, 'Between cure and prosthesis: "good fit" in artificial eardrums' in Jones (ed.), *Rethinking Modern Prostheses*, p. 59.
21 'Optical rays and opinions', *The Optician* (8 November 1892), p. 112. Could possibly be A. Davidson, patent for cork bridge spectacles, patent number 15,928 (6 September 1892).
22 Jones, *Rethinking Modern Prostheses*, pp. 14–16.
23 'A wrinkle in spectacle fitting', *The Optician* (25 March 1897), p. 4.
24 'Notes on novelties and inventions – eye-wire rests for spectacles', *The Optician* (8 April 1897), p. 76.
25 Long, *Spectacles*, p. 23.
26 Christopher Smith Fenner, *Vision: Its Optical Defects, and the Adaption of Spectacles* (London: Lindsay & Blakiston, 1875), pp. 108–9; Carter, *Eyesight: Good and Bad*, p. 246.
27 W.P. Thompson, patent number 9,202 (28 June 1887); G.C. Bateman, patent number 15,300 (26 June 1897); see also discussion of new designs in *The Optician*, for example (25 February 1892), p. 368; (17 May 1894), p. 81; (11 July 1895), p. 176; (17 September 1876), p. 361; (24 November 1899), p. 400; (30 March 1899), p. 48.
28 See, for example, *Liverpool Mercury*, weekly 9 December 1881 until 2 January 1882.
29 'Manchester', *The Optician* (25 February 1892), p. 368.
30 In Science Museum Technical File, T/1921-323: M.W. Dunscombe, *The Evolution of the Bi-Focal Lens* (undated).
31 See, for example, for the cylindrical lens: Rosenthal, *Spectacles and Other Vision Aids*, pp. 246–255; Carl Barck, *The History of Spectacles, Originally Delivered as a Lecture Before the Academy of Science* (Reprinted from the Open Court for April, 1907), p. 15; Richard Corson, *Fashions in Eyeglasses*, 3rd edn (London: Peter Owen, 2011), pp. 132–133; Dorothy M. Turner, 'Thomas Young and the eye on vision', in E.A. Underwood (ed.), *Science, Medicine and History: Essays on the Evolution of Scientific Thought and Medical Practice Written in Honour of Charles Singer* (Oxford: Oxford University Press, 1954), pp. 243–255; Daniel M. Albert, 'Ocular refraction and the development of spectacles', in Daniel M. Albert and Diane D. Edwards (eds), *The History of Ophthalmology* (Oxford: Blackwell Science, 1996), pp. 118–119. For bifocals, see: Rosenthal, *Spectacles and Other Vision Aids*, pp. 258–261; Corson, *Fashions in Eyeglasses*,

pp. 129–133; Wolfgang H. Vogel and Andreas Berke, *Brief History of Vision and Ocular Medicine* (Amsterdam: Wayenborgh Publishers, 2009), pp. 208–209.
32 See, for example, *The Critic* (16 August 1852), p. 435; (15 January 1853), p. 54.
33 See, for example, John Phillips, *Ophthalmic Surgery and Treatment: With Advice on the Use and Abuse of Spectacles* (London: W.B. Keen & Co., 1869), p. 51; Fenner, *Vision*, pp. 108–109.
34 Carter, *Eyesight: Good and Bad*, pp. 252–254.
35 See, for example, 'The Koerts rigid bar eye-glass', *The Optician* (11 July 1895), p. 176; 'Spiller and Martin's improved *pince-nez*' (21 May 1896), p. 138; 'Astigmatic *pince-nez*' (30 July 1896), p. 272; 'A new *pince-nez*' (17 December 1896), p. 176. See also, for example, A.W. Newbold, patent number 23,129 (2 December, 1893); J. Raphael, patent number 8,366 (27 April 1894); G. Spiller, G.S. Martin and A.R. Toole, patent number 9,598 (15 May 1895); C.E. Fitzgerald and G. Prescott, patent number 16,214 (22 July 1896); W. Salt, patent number 8,046 (4 April 1898); R.C. Hines and H.B. Waddey, patent number 16,314 (10 August 1899).
36 Carter, *Eyesight: Good and Bad*, p. 254.
37 'Combined spectacle folders', *The Optician* (20 December 1894), p. 199.
38 Liliane Hilaire-Pérez and Christelle Rabier, 'Self-machinery? Steel trusses and the management of ruptures in eighteenth-century Europe', *Technology and Culture*, 54:3 (2013), 460–502; David M. Turner and Alun Withey, 'Technologies of the body: polite consumption and the correction of deformity in eighteenth-century England', *History: The Journal of the Historical Association*, 99:338 (2014), 775–796.
39 Withey, *Self-Fashioning and Politeness*, p. 92.
40 Science Museum Technical File, T/1921-323, Exhibit of Spectacles, p. 3. However, tortoiseshell was not universally popular, and in 1818 Kitchiner argued that it was too easily broken, and recommended 'silver frames': W. Kitchiner, *Practical Observations on Telescopes, Opera-Glasses and Spectacles*, 3rd edn (London: S. Bagster, 1818), p. 70.
41 See, for example, Alex Faulkner, 'Casing the joint: the material development of artificial hips' in Katherine Ott, David Serlin and Stephen Mihm (eds), *Artificial Parts and Practical Lives: Modern Histories of Prosthetics* (New York: NYU Press, 2002), pp. 199–226.
42 John and Charles Greaves, patent number 1,775 (15 August 1854); Charles Eyland, patent number 451 (22 February 1861).
43 S.Z. Ferranti, patent number 4,584 (8 March 1884).

44 'A visit and description of the house of Messrs Newbold and Bulford', *The Optician* (26 September 1895), p. 13.
45 For America, see Peter John Brownlee, *The Commerce of Vision: Optical Culture and Perception in Antebellum America* (Philadelphia: University of Pennsylvania Press, 2019), pp. 40, 61.
46 Andressen, *Spectacles*, p. 16.
47 Holtman, 'A short history of spectacles', p. xx.
48 In Science Museum Technical File, T/1921-323: Catherine Gates, 'Matthew William Dunscombe: a Bristol optician' (1997), p. 3.
49 Sheffield City Archives, SY231, *Observations on the Choice and Use of Spectacles*, back page.
50 Sheffield City Archives, CA-VAC/119: Sheffield Town Council, Children's Commission Committee, 'From Manufacturers', sheet 2.
51 See, for example, *The Examiner* (27 January 1839), p. 62; *The Belfast News-Letter* (19 October 1841); *The Derby Mercury* (12 December 1877); *Time* (10 July 1887), p. 10.
52 Thos Harris & Son, *A Brief Treatise on the Eyes, Defects of Vision and the Means of Remedying the Same by the Use of Proper Spectacles, Also Rules for judging when Spectacles are necessary, and Directions for selecting them* (London: Onwyhn, 1839), pp. 27–8; William Ackland, *Hints on Spectacles: When to Wear and How to Select Them* (London: Horne & Thornthwaite, 1866), pp. 18–19; Henry Laurance, *The Eye in Health and Disease: With Hints on the Choice and Use of Spectacles*, 3rd edn (London: Love Brothers, 1888), p. 29; Charles Bell Taylor, *How to Select Spectacles in Cases of Long, Short and Weak Sight*, 2nd edn (London: Cassell, 1889), p. 28.
53 See, for example, 'Spectacles and eyeglasses – their forms, mounting and proper adjustments', *The Optician* (15 June 1893), p. 646.
54 Carter, *Eyesight: Good and Bad*, p. 47.
55 'Spectacle glasses', *Chambers's Journal of Popular Literature, Science and Arts* (7 February 1891), pp. 92–93.
56 *Lloyd's Weekly London Newspaper* (11 February 1844); (26 September 1847); (24 October 1847).
57 'British lens maker', *The Optician* (7 May 1891), p. 91.
58 Boots Archives, DA16/8, advance proof of an article titled 'Optics in the applications: some workshop problems and processes', 1890–1900s, p. 2; for more information on Aitchison, see Chapter Four, section headed 'The certificated and qualified dispenser'.
59 Sheffield City Archives, SYCRO 1731, *Progress Commerce, 1893 – Sheffield, Rotherham, and Mexborough* (The London Printing & Engraving Co., 1893), p. 141.

60 London School of Economics, Booth/A/11, Charles Booth's notebook, pp. 78–9. Beyond this it is not easy to track the number of spectacle makers over a given period. In the Census Reports, for example, 'spectacle makers' were listed in 1844, but by 1861 they became 'spectacle makers, opticians' – causing numbers to increase dramatically – and then from 1871 onwards the category of 'spectacle maker' disappeared and only 'philosophical instrument makers, opticians' were individually listed.
61 'A review of Curry and Paxton', *The Optician* (24 January 1874), p. 270.
62 Booth/A/11, pp. 78–80.
63 Ibid., p. 36.
64 Julie Anderson, 'Separating the surgical and commercial: space, prosthetics and the First World War', in Jones (ed.), *Rethinking Modern Prostheses*, pp. 158–178; M. Guyatt, 'Better legs: artificial limbs for British veterans of the First World War', *Journal of Design History*, 14:4 (2001), 307–325.
65 'Frames by Lionel Laurence', *The Optician* (31 March 1898), p. 26; 'Correspondence', *The Optician* (14 April 1898), p. 172.
66 'Correspondence on frame fittings', *The Optician* (21 April 1898), pp. 221–222.
67 Ibid., p. 222.
68 See, for example, comment on the 'Calibration of spectacle lenses', *The Optician* (23 September, 1897), p. 104 and a new stock of lenses 'on the interchangeable plan' in *The Optician* (30 March 1899), p. 116; *The Optician* (16 March 1900), p. 919.
69 'A thirteen stone thirteen year old requires glasses', *The Optician* (20 July 1893), p. 718; 'Spectacles and eyeglasses – III. Prescription of frames', *The Optician* (10 August 1893), p. 768.
70 Boots Archives, DA23/1/18, Aitchison receipt written out to a Master Edwards for some spectacles, 1899.
71 Johann Friedrich Horner, *On Spectacles: Their History and Uses* (London: Balliere, Tindall & Cox, 1887), p. 5.
72 Sheffield City Archives, Bradbury Record 293, *G. and W. Proctor, Opticians and Manufacturers* (Sheffield: C.W. Thompson, 1815).
73 'Popular information on science: loss of sight', *Chambers's Edinburgh Journal* (10 January 1835), pp. 394–395.
74 See, for example, the Science Museum's Dunscombe Collection, a pair of gold eyeglasses 1921-323/299; patent eyeglasses 1921-323/377; patent invisibly cemented bifocal spectacles 1921-323/193; rimless gold spectacles 1921-323/215.

75 These are currently uncatalogued but are available at the Wellcome Library. Catalogues for Stevens' auction house were viewed for 2 March 1915, 13 April 1920, 2 September 1924, 21 and 22 August 1928, 21 September 1928, 14 October 1928, 23 October 1928, 26 and 27 March 1930, 16 and 17 September 1930, 7 November 1930, 5 August 1931, 20 and 21 October 1931, 20 and 21 September 1932, 20 and 21 October 1932 and 16 April 1935; and catalogues for Glendining's auction house for 29 July 1932, 29 October 1934 and 14 January 1935.

76 Asa Briggs, *Victorian Things* (London: B.T. Batsford Ltd, 1998), p. 114; Derek C. Davidson, 'Nineteenth century metal spectacles', *Ophthalmic Antiques International Collectors Club Newsletter*, 58 (1997), 9–10.

77 Data drawn from the accounts books of John Potter Dowell and Robert Sadd & Co: Carlisle Archive Centre, DB9/1-7, John Potter Dowell, Cash and Day Book (Sales), 1885–1898; Cambridge University Library, GBR/0012/Ms Ad.5781–5783, Robert Sadd & Co. Account Books 1837–1851 and Ledgers 1845–1889.

78 Jonathan S. Pointer, 'The £ound in your pocket and the glasses on your nose – 700 years of reading spectacle prices', *Ophthalmic Antiques International Collectors Club Newsletter*, 59 (1997), 4–5.

79 Cambridge University Library, GBR/0012/Ms Ad.5781–5783, Robert Sadd & Co. Account Books 1837–1851 and Ledgers 1845–1889.

80 See, for example, newspapers: *Examiner* (21 January 1838); *The Morning Chronicle* (7 July 1840); *The Bristol Mercury*, (22 October 1853); *Ipswich Journal* (3 June 1854) and weekly between 7 December and 21 December 1895; *Caledonian Mercury* (30 April 1855); *Belfast News-Letter* (10 October 1865); *Cheshire Observer and Chester, Birkenhead, Crewe and North Wales Times* (22 December 1866); *The Bristol Mercury*, weekly between 16 June 1877 and 21 July 1877; *Freeman's Journal and Daily Commercial Advertiser*, weekly between 22 October 1879 and 5 November 1879; *Nottinghamshire Guardian* (7 November 1879); *Glasgow Herald* (23 January 1883); *The Dundee Courier & Argus*, weekly between 15 September 1883 and 8 December 1883; *Leicester Chronicle and Leicestershire Mercury*, weekly between 22 March 1884 and 7 June 1884; *The Sheffield & Rotherham Independent* (11 October 1887); *Glasgow Herald* (26 February 1889). For other periodicals, see: *The Athenaeum*, weekly between 4 January 1851, p. 1 and 29 October 1853, p. 1302; *Longman's Magazine* (April 1893), p. 655.

81 *The London Reader: of Literature, Science, Art and General Information*, regularly between 28 November 1868, p. 121 and 27 August 1870, p. 409; *The Leicester Chronicle and Leicestershire Mercury* (7 January 1888), p. 4; and weekly between 18 February and 7 April 1888.
82 *The Wrexham Advertiser, and North Wales News* (26 March 1887); *Belfast News-Letter* (10 October 1865); *Glasgow Herald* (26 July 1871) and (3 August 1871).
83 Heather R. Perry, 'Re-arming the disabled veterans: artificially rebuilding state and society in World War One Germany', in Ott, Serlin and Mihm (eds), *Artificial Parts and Practical Lives*, pp. 75–102. This is not always the case; see Ryan Sweet's argument that appearance was important for lower-class users: *Prosthetic Body Parts in Nineteenth Century Literature and Culture* (London: Palgrave Macmillan, 2022), p. 179. Open access: DOI:10.1007/978-3-030-78589-5.
84 *The Bristol Mercury* (21 March 1846); *The Dundee Courier & Argus*, weekly between 30 June 1883 and 15 March 1884.
85 'Spectacle secrets', *Tait's Edinburgh Magazine* (December 1838), pp. 803–804; 'Electric spectacles', *The Standard* (23 August 1893); John Grimshaw, *Eyestrain and Eyesight; How to Help the Eye and Save the Sight* (London: J. & A. Churchill, 1907), p. 40.
86 Jones (ed.), *Rethinking Modern Prostheses*, p. 4.
87 The National Archives, MH 12/11000/19, Spectacles for aged inmates, 4 April 1853, folio 47; The National Archives MH 12/9534/77, From the [sic] Harry Farnall, poor law inspector of Southwell workhouse, 30 October 1867, folios 105–107.
88 'Short sighted children', *The Standard* (22 December 1888).
89 Carlisle Archive Centre, John Potter Dowell, DB9/2, Cash day book (sales) 1885–1886, Day book at the back, 20 March 1886; DB9/4 Cash day book (sales) 1888–1890, 11 July 1888, 17 April 1888, 18 October 1888; DB9/5 Cash day book (sales) 1891–1892, Day book at the back, 11 June 1891, 26 April 1892, 5 May 1892; DB9/6 Cash day book (sales) 1893–1894, 16 November 1893, 2 July 1894, 24 November 1894 [unpaginated].
90 See, for example, draper Mr J. Irving Bell in DB9/5 Cash day book (sales) 1891–1892, Day book at the back, 9 July 1891; DB9/6 Cash day book (sales) 1893–1894, 4 April 1894; ironmonger Mr J.G. Parker in DB9/2 Cash day book (sales) 1885–1886, Day book at the back, 6 July 1885, 16 January 1886, 6 February 1886 [unpaginated].
91 Cambridge University Library, GBR/0012/Ms Ad. 5783, Robert Sadd & Co. Ledger 2., p. 111.

92 See, for example, 'Optical queries and answers', *The Optician* (15 July 1897), pp. 298–299; (9 December 1897), p. 320; (6 October 1898), p. 188.
93 London Metropolitan Archives, A/KE/B/01/04/004, Royal London Ophthalmic Hospitals, Complaint that there was little profit in supplying spectacles, letter dated 18 October 1900.
94 'Eyesight Good and Bad', *The Morning Post* (12 February 1880).
95 'Health and Education Department', *The Leeds Mercury* (5 October 1883).
96 'Multiple news items', *The Morning Post* (18 September 1884), p. 4.
97 'The modern spectacle maker – a visit to Culvers', *The Optician* (27 September 1894), p. 6.
98 Sheffield City Archives, *Sheffield, Rotherham, and Mexborough*, p. 141.
99 London Metropolitan Archives, A/FWA/C/D/223/001, London Spectacle Mission correspondence and papers, private and confidential report, 1 January 1896.
100 London Metropolitan Archives, A/FWA/C/D/223/001: London Spectacle Mission correspondence and papers, private and confidential report, 4 January 1896.
101 London Metropolitan Archives, A/FWA/C/D/223/001, London Spectacle Mission correspondence and papers, private and confidential report, 8 November 1899 and 26 May 1900.
102 London Metropolitan Archives, A/FWA/C/D/223/001, London Spectacle Mission correspondence and papers, pamphlet *Spectacles for the Needy, and How to Get Them*, by Sibyl Bristowe, undated [but possibly 1912].
103 London Metropolitan Archives, *Spectacles for the Needy*; London Metropolitan Archives, A/FWA/C/D/223/001, London Spectacle Mission correspondence and papers, newspaper clipping from *Truth*, 8 May 1902.

4

The limits of professionalism: medical practitioners, opticians and popular responses to sight loss, 1880–1904

On 14 December 1894 an anonymous patient wrote to the *Aberdeen Weekly Journal* to describe their experiences at the Ophthalmic Institution, situated on the site of the Aberdeen General Dispensary.[1] A straightforward reading of this account documents the ascent of the medical practitioner in public responses to problems with their vision and eye health. The patient attended the institution to consult an 'eye doctor' for short-sightedness, having sought the advice of a 'medical friend'. The ophthalmic surgeon, Dr Mackenzie Davidson, was described as a 'clever specialist' and a respected public figure. Amongst the working classes, Davidson's ability had a mysterious, miracle-like, quality, to which the patient argued that 'such is the faith… they would not be very much astonished if he were to restore the sight of another Blind Bartimeus by a simple wave of the hand'. The patient observed, too, that medical students 'listened to him with great deference' and that Davidson spoke and moved 'with the air of one that knows all about it' as he preceded to test and examine their vision using optotypes 'twenty-five feet off' and a 'dummy pair of spectacles' to trial lenses of various powers. Following the administration of drops, the patient then had their eyes examined using an ophthalmoscope, which they found to be both 'mysterious' and 'fatiguing'. On the completion of the ophthalmoscopic examination, the patient was given a prescription, advised that they had astigmatism and told that they could purchase spectacles at a nearby optician's firm 'at a cheap rate to patients from the Ophthalmic Institution'.

The Ophthalmic Institution, in a poorer part of the city, served the working-class communities of Aberdeen. The building was not on an 'aristocratic site' and did 'not meet the local fashionable eye', and therefore the patient was writing both to describe their experiences and encourage donations to an institution that depended 'mainly on voluntary annual or occasional contributions'. Both the character of Davidson – his kindness, his philanthropy and his work as a 'practical Christian' – and his skill as an ophthalmologist were deemed important to the patient in presenting their case to the newspaper's readers. It is immediate, and should therefore be borne in mind, that the writer was not a typical patient of the dispensary. They did not identify with the 'great bulk' of patients that were 'wretchedly clad', and the patient understood the terms 'ophthalmoscope' and 'myopia', of which the latter caused the ophthalmologist's assistant to look at them 'in such a way that I infer first, that the patients there are seldom or never able to name their own ailment'. How typical was this patient's experience and/or knowledge of vision outside the working-class communities of Aberdeen? In the early 1890s, did the dispensing of spectacles typically involve a vision test in a hospital before sale on the high street through established relationships between opticians' firms and hospitals? Was an ophthalmologist now considered to be the cultural authority on matters of vision testing and spectacle use?

By drawing upon both medical and trade journals, medical trade catalogues, advertisements, pamphlets, private correspondence and objects in the Science Museum's collections, I argue in this chapter that medical practitioners did not achieve either an ideological or a practical monopoly over vision testing and subsequent spectacle dispensing. Instead, they navigated an ever more divaricate commercial market than the one discussed in Chapter One, a market that held a position different from that of other prostheses and assistive technologies in the Victorian medical marketplace. As already argued by Claire L. Jones, traditional approaches to the medical marketplace have not adequately foregrounded the medical practitioner as either a producer or a consumer of medical products.[2] Takahiro Ueyama has similarly demonstrated that medical practitioners in late Victorian London were forced to engage and establish relationships with traders because of the expanding commercial market for medical products and a growing health culture among the general public. Rather than 'division' and 'conflict', Ueyama points to 'entanglement' and 'collaboration'.[3] Between 1880 and the early twentieth century this expanding medical market encompassed patent medicines, pharmaceutical remedies and a variety of technological or assistive devices intended to improve a person's health or alter their physical or sensory capacity, often to help them conform. In the case of prostheses, Julie Anderson and Caroline Lieffers have shown that makers of artificial limbs and medical practitioners' practices were similarly intertwined, and relationships could be both collaborative and effective.[4] In a comparable vein, it would be simplistic to ascribe spectacle sellers' and medical practitioners' actions in the 1890s to active conflict; in the heat of debate there were shared concerns and calls for cooperation.

Nevertheless, in more recent scholarly work the commodification of prostheses has been seen as a vital part of asserting medical control over people's bodies, and this is where a study of spectacles departs from that of other bodily devices.[5] Although direct comparisons have been made between spectacles and these devices in the eighteenth century, by the late Victorian period spectacles occupied their own market and operated in this market differently to, for example, the steel truss, the artificial limb or the hearing aid.[6] The variety of purposes for spectacles – to aid reading in old age, to treat more complex refractive and accommodative eye conditions – meant that, unlike other prostheses, arguably spectacles could be, and arguably they were, mass-marketed. As we saw in Chapter Three, by 1880, when ophthalmologists were entering the market more forcibly, the production of spectacles was already vast, visible and diverse. Patrick Wallis has argued that a focus on services as opposed to goods in the early modern medical economy offers a new insight into both medical and popular consumption.[7] In the case of spectacles, too, it is important to note a shift by the 1880s from the marketing of goods – spectacles – to that of a service – the testing of vision and adapting a frame to a person's face. This had professional implications for spectacle dispensers and medical practice; the dispensing transaction had altered from a more conventional commercial transaction to one that depended on the judgement of a dispenser in a position of trust.[8] Ophthalmologists certainly influenced this marketplace, and we will see, for example, how opticians and traders continued to alter their practice and the language of their advertisements. However, medical practitioners also felt threatened and did not control their practical working relationships with retailers. Unlike their early Victorian counterparts, opticians now emphasised their ability to complete ophthalmologist's prescriptions and establish medical partnerships, on one hand, and operate the ophthalmoscope and conduct vision tests, on the other. In contrast to the way the physical distance between artificial limb makers and medical practitioners was reduced during the First World War, through the establishment of rehabilitation centres, the blurring of expertise in vision testing between spectacle sellers and the medical faculty necessitated a separate working partnership of effective communication.[9]

The commodification of spectacles and alterations in the dispensing process challenged the role of the medical practitioner in the dispensing of visual aids, rather than asserting it. The marketing of spectacles between 1880 and 1901 aided in the standardisation of visual capacity while simultaneously challenging that standardisation through the number of spectacles retailed in a variety of different contexts. The retailing optician of the 1880s and 1890s varied. My use of the term 'optician' does not include just the optical or scientific instrument maker. Like its late Victorian usage, my use of the term 'optician' identifies an individual who studied, and specialised in, either optical goods, or the measurement of vision and treatment of simple refractive vision errors. It therefore encompasses a variety of chemists, jewellers and watchmakers as well as traditional opticians. In all the term's usages there remained two distinct groups: the trader that simply fitted or constructed frames and lenses, and an early form of optometrist that adopted ophthalmologists' vision testing techniques.

The intensity of debate among the latter, other traders who did not wish to specialise and ophthalmologists reflects the professional implications of gaining a monopoly on vision testing. Control of the retailing market for spectacles would consolidate ophthalmologists' position as a public expert and overseer of the ocular health of the British nation, aspirations that were discussed in Chapter Two. Also, we should not understate the potential this position had to be a lucrative commercial arm to ophthalmologists' more widely projected philanthropic practice. But opticians, like other rising professions in the nineteenth century such as bonesetters and druggists, challenged the mobile boundaries of medical orthodoxy.[10] This undermining, and the debates that ensued, were of practical importance both professionally and economically, and involved multiple actors. While medical ideas certainly transformed the market for and retailing of spectacles, medical practitioners' attempts to gain a monopoly were unsuccessful. The advice of an ophthalmologist or a visit to a hospital was not always a person's first attempt to get treatment. The title 'optician', and its traditional role in spectacle dispensing, continued to hold considerable authority, as too did the idea that spectacles could be bought through a simple commercial transaction. Assistive devices are not made, adapted and sold in a vacuum. Users are important in determining outcomes of use and, in this case, Victorians were reluctant to view spectacles through a solely medical lens or be fully impressed with alterations in vision testing. For many, spectacles remained a device that could be bought on the high street whenever the patient or customer felt that their vision was failing and not based on advice dictated by an expert, whether medical or optical.

Locating the late Victorian spectacle retail market

A case of trial lenses made by C.W. Dixey and owned by British ophthalmic surgeon Sir Anderson Critchett can be found in the Science Museum's Ophthalmology collection. Critchett retired in 1903 after twenty years of service at St Mary's Hospital to become Surgeon Oculist to King Edward VII.[11]

In the obituary published following his death in 1925, Critchett was described as having been 'the doyen of ophthalmic surgeons' and had a strong pedigree – his father was one of the founders of modern ophthalmology. Critchett was considered to be one of the best-known British surgeons in the ophthalmic world.[12] While the case itself cannot be accurately dated, Critchett began his practice at St Mary's Hospital in the early 1880s. The case embodies a professional relationship between the ophthalmologist, using a

Figure 4.1 A case for trial lenses owned by Sir Anderson Critchett, complete with lenses and the coat of arms of C.W. Dixey printed on the inside cushioning. Undated.

Source: Science Museum Ophthalmology collection, A635437. With permission from SSPL/Science Museum.

set of trial lenses to test a person's vision, and the optician who had manufactured the necessary equipment. Specialist tools were often designed and produced by practitioners working in partnership with instrument makers in a system governed informally by personal relationships.[13] Between 1875 and 1903, the *British Medical*

Journal provides evidence for working partnerships between several opticians and medical practitioners in the production and improvement of new sight-testing technologies. Mr Percy Dunn, FRCS, for example, ophthalmic surgeon to the West Hospital, stated that he was 'much indebted for the skill and care' shown by F. Davidson, a London optician at 140 Great Portland Street, in the creation of a model eye to train students in retinoscopy and the use of the ophthalmoscope.[14]

These small insights into medical practitioners' and opticians' working relationships, however, suggest that the dynamic of that relationship was unequal. While ophthalmologists were indeed 'indebted' to the specialised skillset of the optician, Newcastle optician Mr Robson produced a new optometer 'under the supervision' of a medical practitioner, Christopher Jeaffreson.[15] As well as developing new sight-testing technologies, medical practitioners also believed that the dispensing and retailing of spectacles should be overseen by medical experts (themselves). In 1862 Joel Soelberg Wells proposed the early adoption of such a plan, a plan that he claimed was already in use on the Continent and by 'several ophthalmologists' in England:

> The medical man himself selects the proper glass from his spectacle box (which contains concave and convex glasses, corresponding number[s] being kept by the optician); the focal distance of the required glass is written on a slip of paper, which is taken to the optician who supplies the patient with the spectacles prescribed thereon. Thus we are sure that the patient is furnished with proper glasses.[16]

Yet by 1880 significant changes had taken place and the high street was no longer the only site for the supply of spectacles, as it had been fifty years before (see Chapter One). The cashbook subscriptions to the West of England Eye Infirmary show that London wholesale opticians Pickard and Curry – from October 1889, renamed Curry & Paxton – had a new customer: the hospital, which had sought a regular supply of spectacles between 1885 and 1895.[17]

Medical trade catalogues grew exponentially between 1870 and 1914 and were a well-placed form of print culture that medical personnel could utilise to obtain new diagnostic technologies without consulting the optician on the high street. Far from being a step outside the commercial world, medical trade catalogues were one of the most prominent forms of advertising targeting medical practitioners, in the guise of a medical publication and selling tools vital for medical practice. As Jones argued, the catalogue was 'a printed embodiment of the compatibility of the medical profession and commerce'.[18] In the Thackeray Museum's collection of medical trade catalogues fifty-five examples included vision testing equipment and/or spectacles. As discussed in Chapter Two, catalogues' extensive lists and illustrations are a visual insight into the expansion in the number and types of diagnostic equipment relating to vision testing. In 1863, shortly after the invention of the ophthalmoscope and the publication of Donders, London-based company James Weiss & Son included in their catalogue a single ophthalmoscope and trial lens case.[19] In contrast, consecutive runs of catalogues for individual companies such as London-based Down Bros between 1885 and 1901 reveal a steady growth, and the inclusion of trial lenses, ophthalmoscopes and test charts by 1901 to equip practitioners with the means to test people's vision more thoroughly while bypassing the optician.[20]

The medical trade catalogue offered a further opportunity to physically distance the practice of the optician from the medical faculty by internally controlling the retail and supply of spectacles.[21] The catalogue ensured compliance in a medical environment that sought to restrict advertising, and companies targeted medical specialists that tended to incorporate, and embrace, a range of new tools.[22] Medical trade catalogues for companies, based both in and outside London, provided an alternative system for prescribing spectacles that bypassed the optician or retailer on the high street. From the 1880s, firms in Birmingham, Leeds, London, Manchester and Newcastle advertised spectacles and visual aids. These surgical instrument and medical appliance manufacturers mimicked, and could act as a substitute for, the relationship between the optician and an ophthalmic practitioner or institution. In 1887, for example, Reynolds & Branson of Leeds stated that they 'devote special attention to the prescription of Surgeon oculists, all glasses being carefully tested previously to dispatch'.[23] The catalogue itself acted as a tool for asserting the medical practitioner's role in the dispensing of spectacles. In 1897 Brady & Martin of Newcastle, for example, described the increasing desire of medical practitioners to monopolise the retail and supply of visual aids:

> The careful selection of spectacles... is now universally recognised. Most Medical Practitioners will prefer to order the particular spectacles required by their patients.[24]

This new body of traders saw the commercial potential of medical practitioners as consumers of medical products, a trade which included spectacles. Their practices straddled the retailing optician on the high street and blurred the boundaries of opticians' and medical practice, an overlap that can be seen in the newspaper advertisements of Lindsey and Sons. This firm specialised in steel and presented themselves to the readers of *Jackson's Oxford Journal* as surgical instrument makers, cutlers and opticians.[25] Every week between 22 September 1888 and 7 September 1889 they emphasised that they were 'agents for Messrs Pickard and Curry's Spectacles'. On 7 September 1889, however, while continuing to be agents for London opticians Pickard and Curry, they also introduced 'oculists prescriptions carefully worked'. While they may have manufactured their own, or sourced different frames, Lindsey and Sons could conceivably have used Pickard and Curry's frames and lenses in their partnership with oculists.

The meaning and role of an optician altered from the 1850s in response to changes to the vision test and newfound sites of commercial sale to which Pickard and Curry had responded: the hospital and surgical instrument maker. A case containing a pair of straight spectacles in the Science Museum's Ophthalmology collections is inscribed 'Priest & Ashmore, Ophthalmic Optician, Sheffield'.[26] Priest & Ashmore's trade directory entries demonstrate the influence of medical practice on the firm's name and the need to promote their practice on the high street as respectable. Their name evolved from manufacturing opticians in 1833, to add 'makers of ophthalmologist instruments' and eventually adopt 'oculists' prescription work' as a specialty in 1901.[27] Such phrasing aligned opticians with a role similar to the makers of surgical instruments and medical appliances in their medical trade catalogues, and the similarities of language are striking. In 1895 a Scottish 'Chemist and Optician', Henry Milne advertised in the *Aberdeen Weekly*

Journal 'oculist's prescriptions accurately prepared'.[28] Similar to Milne, a range of opticians consciously created working partnerships with ophthalmologists that often reduced their role to only working with, and providing spectacles from prescriptions supplied by, ophthalmologists. In these instances, opticians marketed and promoted their 'goods' and focused on their ability to accurately work lenses and fit frames to a person's face.

Other opticians were less conservative; they altered their practice and promoted a vision testing service. L.J. Troulan & Son, for example, chose the term 'oculistic opticians' on their leather frog-mouth case.[29] By the 1880s several opticians' firms advertised on their wares and in newspapers a name – such as 'ophthalmic optician' or 'oculistic optician' – that sought to closely align their practice with that of the emerging ophthalmologist, a practice that, in both instances, incorporated the more sophisticated vision test. This was different to claiming an association with a medical institution or personnel, or the carrying out of oculists' prescriptions, that had existed and evolved in opticians' advertisements from the 1830s. The content of these late Victorian advertisements alludes to practical changes in the way that they dispensed spectacles. To say that their dispensing was 'accurate' or 'scientific' in the 1880s was a fundamentally different claim to that of their peers in the 1830s. Lenses measured in the dioptre required a greater degree of precision, and achieving that precision increasingly required the use of testing equipment that opticians had manufactured and advertised since the devices were invented in the 1850s. In the Science Museum's Ophthalmology collection, handling the equipment that opticians both manufactured and used offers a tangible insight into the way that those at the forefront of their practice were forced to change. Diagnostic technologies were no longer a small set of five test spectacles that could be placed in the pocket. The new trial lens cases were large, engraved wooden boxes, with diverse lens options, and not especially portable. Indeed, in 1890 a Manchester optician, A. Franks, advertised that his spectacles were 'scientifically tested' and could be 'adapted' to his customers' eyesight in his specific 'eyesight testing room'.[30] In Liverpool, in 1895, optician 'Wood, late Abraham' advertised the firm's change of location and the opening of 'new eye-testing rooms'. Wood sought new premises because 'much space is needed' when 'persons having their sight tested have to sit about 16ft from the test-types' and 'when several sights are tested at one time'.[31] Both Wood and Franks not only advertised that vision testing should be carried out in a controlled environment, but that they, the dispenser, could test and adapt optical lenses to suit their customers' vision.

Scholars have argued that the use of equipment by medical practitioners – such as the stethoscope and thermometer – increased medical authority, and patients' subsequent passivity, in the diagnosis of illness.[32] In 1880 prominent English ophthalmologist Robert Brudenell Carter argued that patients should be passive during the initial diagnosis of the refractive state of their eye, because their impressions were 'untrustworthy' and people with 'half vision' often believed that their vision was 'remarkable'.[33] The introduction of technology into the process of dispensing spectacles crucially created a new position of expertise, one that shifted knowledge away from the customer or patient's experience of their vision and the determination of the lens that would best suit them to the dispenser, a position to which both ophthalmologists and opticians laid claim. Several historians have analysed the methods used by medical practitioners to either subjugate or completely remove occupations that straddled the borders of their practice, such as nursing, dispensing pharmaceuticals, dentistry and orthodontics.[34] In such studies, historians have used the term 'professionalisation' loosely, to describe the definition of a distinct occupational group with a particular knowledge base.[35] These are often internal accounts focusing on definitions, the establishment of journals, education, certification, a society and a code of ethics. A fundamental feature of professionalisation often missed is its interdependent nature: it is controlled, even defined, by jurisdictional boundaries; indeed, a profession emerges when there is a vacancy and/or an occupational group seeks to enlarge its own jurisdiction.[36] The actions of ophthalmologists and retailing opticians thus offer crucial new insights into this area. Ophthalmologists acted from a far from secure, even vulnerable, position, having only established their own formal society in 1880 and with little training in vision testing or refractive vision errors in a medical curriculum that supported a more generalised form of training.[37] As Jones has argued, medical specialists adopted new technologies as a way of challenging the methodology of the more elite physician.[38] I argue that, for ophthalmologists, the adoption of technology meant that they needed to confront the trade market from which it came.

As we saw in Chapter Two, it was from the 1880s that ophthalmologists sought to consolidate their position as public experts. But they also needed to make a living.[39] Ophthalmologists understood the lucrative potential of spectacle sales in addition to diagnosing and treating ophthalmic disease. While they had cultivated a relationship with the state on matters of vision in schools and the workplace, the retailing market for spectacles posed a far greater challenge. By the 1880s the optician had a long association with spectacle dispensing – an association that extended over centuries, not decades – and one that persisted, and they aspired to maintain, regardless of the seismic shifts in vision testing since the 1850s. The role of opticians in spectacle dispensing is an important example for challenging the assumption that the commodification of prostheses entrenched the medical model of disability and medical influence.[40] For the optician, this was not a straightforward matter of pitching their case against ophthalmologists' practices in a bid to professionalise their area of expertise and continue vision testing. Opticians were required to position themselves between medical practice and a vast retail market that encompassed a greater variety of different traders than the early Victorian market explored in Chapter One. Launched on 2 April 1891 the trade's journal, *The Optician*, sought to bring together opticians who still tested vision and laid claim to the practice. Its first leader called for greater knowledge and training for practising opticians and a standardised, cohesive community that could demarcate their practice from that of other traders and, simultaneously, 'wipe out' medical practitioners' 'covert sneer' contained in the epithet 'Shoptician'.[41] Two months later an optician under the pseudonym 'Fair Play' explained that:

> [t]here are Opticians and Opticians, and there is just as much difference between a spectacle dealer and a first-class Optician as there is between a dabbler in oculistic matters and a qualified surgeon oculist.[42]

Who was the 'spectacle-dealing' optician in the early 1890s? In the early issues of *The Optician*, its editors and correspondents created a character that 'prowled' the streets or 'filched' business from their high street stores. Indeed, newspaper advertisements for spectacles in 'THE BEST PLACE' and 'for half the usual price' suggest little change had taken place in spectacle dispensing practices since the mid-century.[43] Continuity is certainly evident in the still-active relationship between the jewellery trades and the dispensing of spectacles, such as in the advertisements of a 'Silversmith and Optician' at 'High-Street on Gravesend' in 1852 and of a 'Goldsmith and Optician' in the *Northern Echo* in the 1890s.[44] The high street market for spectacles, however, had expanded vastly since the 1850s and there were several newcomers. In the early 1890s *The Optician* cautioned against the encroachments of chemists, druggists, stationers and drapers on the 'traditional right' of the optician and optical instrument maker.

Newspaper advertisements document the reality of *The Optician*'s perceived problem and the potential difficulty of identifying a reputable seller. Ueyama has outlined the advertising strategies of a newly modernised and 'scientific' form of the quack retailer that continued to persist but in 'new garb'.[45] By the 1890s the optician was an ambiguous category of trader and one with variable vision testing and optical expertise. 'Oculist Optician' Henry Laurance, for example, presented to the public a convincing case of a 'first-class optician'; he advertised three branches in Birmingham, Manchester and Glasgow, as well as his pamphlet *The Eye in Health and Disease, with Hints on the Choice and Use of Spectacles*, which was in its 3rd edition in 1888. In his pamphlet Laurance advised the public on the states of 'normal' and 'abnormal' vision, the 'defects of vision rectified by spectacles alone', 'cases in which medical or surgical treatment is needed in addition to spectacles' and how to choose and order spectacle frames.[46] He carefully positioned his practice alongside that of the ophthalmologist and warned against the 'haphazard plan of selecting spectacles employed by some opticians'. Despite this, Laurance's own conduct and advertising strategies were sometimes questionable. He proposed that vision could be tested via post when a 'personal' interview' was not possible, a practice that *The Optician* and medical practitioners now condemned as out-dated. Moreover, in addition to his three branches, Laurance had various 'agents' for his spectacles, agents who claimed a legitimate association with his name but whose own vision testing expertise and/or knowledge as jewellers, cutlers and chemists was uncertain.[47] The adoption of the title 'optician' by these traders – 'jewellers and optician', 'cutler and optician' and 'chemist optician' – demonstrates a shift in the meaning of the term optician to signify an individual who dispensed spectacles and eyeglasses, and not necessarily optical instruments; a trader who could have little expertise in either the physiological state of the eye or the optical laws of light refraction.

Laurence's pamphlet targeted a growing popular demand for information about eye care and the use of spectacles. Jokes in weekly periodicals indicate that 'myopia' and the medical diagnosis of vision were part of a broader range of complex technical jargon that alienated popular and elite understanding. A much earlier example of this, taken from *The New York Herald* and printed in *The Belfast News-Letter* in 1863, revealed a patient's inability to interpret their surgeon's diagnosis:

> 'Young man', said the surgeon, looking me straight in the eye, 'you have got the myopia'. 'Yes, Sir', said I, 'and good ones, too – a little Binniger, with a drop of Stoughton makes an excellent eye opener of a morning'. 'And there seems to be an amaurotic tendency of the right eye, accompanied with ophthalmia.' 'Show', says I. 'And that white sort on the left eye betokens a cataract.' 'I guess you mean in the ear', says I, "cause I went in swimming this mornin', and I got an all fired big bubble in my left ear', and here I jumped up two or three times on my left foot, but to no purpose.[48]

Newspapers therefore addressed this problem by describing myopia using alternative and more familiar terms. The columns 'Scraps' in the *Manchester Times* in 1882 and '*Factae*' in *The Royal Cornwall Gazette* in 1883 impressed on the public the deterioration of vision in Germany by arguing that 'myopia, which is doctor for "short sightedness" is spreading with startling rapidity in the German schools'.[49] Ophthalmologists reinforced this divide between elite and popular levels of understanding to define their area of authority and expertise as texts in 'clear and simple' language simultaneously circulated in the 1880s. George Black, MB Edinburgh, for example, argued that his text would offer advice on 'eye defects' and vision that were often a 'practically sealed' book.[50]

Opticians used the pamphlet as a similar tool to educate the public and promote their own practice and expertise by providing practical information on spectacles and their use. Crucially, as Ueyama demonstrates, traders were not relying on advertising strategies alone, but were tapping into the health-conscious Victorian's psyche. Like the broader medical marketplace, traders adopted advertising strategies that capitalised upon a rapidly growing public demand for preventative measures to preserve health in the 1880s.[51] London optician John Browning's 1887 publication *Our Eyes and How to Preserve Them from Infancy to Old Age*, for example, was as much a guide to preserving eye health as it was a guide to the selection of appropriate spectacle frames.[52] Yet, as a strategy adopted by opticians much earlier and discussed in Chapter One, by the 1880s opticians appeared authoritative and had produced several editions of their works: Browning's *Our Eyes and How to Preserve Them* was in its seventh edition. Not surprisingly therefore, opticians' publications were met with antipathy by medical practitioners. In 1886, for example, the *British Medical Journal* printed a scathing review of Browning's work, arguing that a 'patient who buys glasses at an optician's does so at his own risk'.[53] However, their concerns in the 1880s about any challenge the optician might have presented may have been premature. In 1889, J. Gray Keith, 'formerly assistant optician to the queen and late science lecturer to Queen's College', offered a 'reliable guide' on the eyesight, use of spectacles and care of the eyes, with five cautions.[54] In the title of his publication, Keith argued that thousands 'ruin their sight', or 'are rendered blind', from failing to heed opticians' advice. Such rhetoric would have had considerable persuasive sway for a newly health-conscious public especially when, as discussed in Chapter Two, concerns over the deterioration of vision and the increase in blindness in Britain and the Western world hit the headlines of newspapers from the 1880s. Yet it also hints at the number of individuals not attending a 'thoroughly skilled optician', wearing cheap spectacles, or avoiding the use of spectacles altogether:

> **Don't** wear the cheap common glasses...
> **Don't** try to suit yourself with a pair ready-made...
> **Don't** try to do without spectacles if your sight is failing...
> **Don't** hesitate to apply to a thoroughly skilled optician...
> **Don't** apply to a mere vendor of spectacles, or to a watchmaker, jeweller, stationer, chemist, ironmaker &c. or indeed any person who does not make *special* study and business of adapting spectacles and eyeglasses.[55]

It would be anachronistic to assume that the optician or medical practitioner would be a person's first point of call. As Anne Digby has suggested, 'self-help' ideologies continued to persist in the second half of the nineteenth century, despite the parallel growing profile of, and confidence in, the medical profession.[56] For eye care and spectacles, 'self-help' could mean resorting to medical practitioners' and opticians' popular texts or, as these texts decry, avoiding any who claimed expertise in vision testing and spectacle dispensing, as such *laissez-faire* attitudes conflicted with the idea of providing a professional vision testing service.[57] A glance through the advice columns in popular newspapers and periodicals reveals that people might turn to a weekly periodical when initially concerned about their vision. In 1890 a correspondent from Salford wrote to the London Saturday periodical *Bow Bells* to ask 'I am very near-sighted. What can I do besides wearing spectacles to improve my eyesight?'[58] The periodical advised the correspondent to 'consult an experienced oculist', an option the correspondent had been unaware of prior to writing or chosen to avoid, either because they questioned the practice or were uncertain who might be competent to consult. In these instances, both newspapers and periodicals recognised an opportunity to present to the public well-informed information about vision and how to care for it. In addition to health, there was a widespread public interest in many branches of science in this period, a combination of interests for which spectacles were well positioned.[59] Newspapers and periodicals produced their own articles – such as 'familiar talks about physiology' – or published findings from recent papers or lectures. The information circulating in popular literature supported medical practitioners' and opticians' attempts to reform spectacle dispensing practices and did little to challenge their stance on spectacle dispensing. The newspapers, in particular, were more concerned with the quack, a problem that they had long targeted because of unregulated advertising.[60] In 1890, for example, in the column 'Chats with housekeeper' in the *Newcastle Weekly Courant* Phyllis Browne argued that 'nothing can be more foolish than for a person to go into a shop, "try on" a few spectacles, and then purchase the pair through which they can see best'.[61] In the same issue, the newspaper's advertising was more conservative and only included Brady & Martin, scientific instrument makers of 27, 29 and 31 Mosley Street, Newcastle-on-Tyne, who advertised spectacles made using the 'best glass' alongside their opera glasses and microscopes.

Correspondents writing to a newspaper to air concerns over failing eyesight offer a small glimpse into what scholars have argued to be a questioning and careful – not naïve and passive – consumer. By the 1880s and early 1890s such consumers, while sceptical and savvy, were confronted by an overwhelming volume of traders both in the advertising columns of popular literature and the high street that adopted scientific language and sought to present a respectable and professional image.[62] Their navigation of the spectacle market was further complicated by the inability to identify legitimate traders dispensing spectacles by trade name. Correspondence by an 'Insider' in *The Optician* in 1892, under the heading 'a serious matter for the public', argued that 'almost daily one meets with advertisements from all sorts and conditions of men, in all sorts of trades, who are open to supply anything and everything'.[63] Insider's solution was to encourage opticians to 'help the buyer discriminate' and emphasise the permanent injury incurred using the wrong lenses.

While several traders did tag either the sale of spectacles to their stocklist or the title 'optician' to their firm's name, as Insider suggests, it was not, as he proposed, a straightforward hierarchy of sale on the high street with the optician and optical instrument maker at the top and other 'encroaching' trades at the bottom. An 'optician' might have no background in optical laws or spectacle dispensing, while a chemist or jeweller might be well qualified in both. Thomas Armstrong & Brother of Manchester, established in 1825, advertised prolifically in the latter half of the nineteenth century in the *Manchester Times*, as well as *The City Jackdaw* and the *British Architect*. The firm's trade directory entries between 1879 and the 1890s stated that they were 'opticians by appointment to the Eye Hospital', as well as 'watch and clock manufacturers and importers, jewellers, silversmiths… and mathematical and philosophical instrument makers'.[64] Armstrong & Brother did not specialise entirely in the sale and dispensing of spectacles. However, their relationship with the Manchester Eye Hospital was more than an advertising ploy. An entry in the diary of Edward M. Wrench, a Derbyshire surgeon, for example, detailed that he took 'Kirlee' to Dr Little of Manchester to obtain a prescription for spectacles on 11 June 1883. On the prescription ticket, also dated 11 June 1883 and enclosed in the diary, Thomas Armstrong & Brother were asked to 'please supply Master Wrench, D *plane glass*… S +24 pebble'.[65] This rare survival of ephemera, a prescription ticket, indicates that Armstrong & Brother were engaged in prescription work and deemed by medical practitioners a reputable place for their clients to obtain, and be fitted with, spectacles. It also indicates how the practical working relationship between a retailer and medical practitioner – and its advertisement – offered a reassuring validation of the quality of their products to the patient customer.[66]

Accounts by collectors of visual aids and scholars focusing on the history of optometry and ophthalmology have argued that there was a geographical divide in spectacle dispensing. In Chapters One and Three I challenged this stance by arguing that the relationship between the type of sale and/or trader and the location of sale is more complex. Just as Chapter Three showed that spectacles could be obtained in major cities and provincial towns, advancements in spectacle dispensing practices and changes to traders' advertising were not isolated to London. Certainly, hawkers were still attempting to make a living from selling spectacles on the street in provincial areas. Correspondence relating to deaf German migrant Wilhem Bauman in Cambridge's charity organisation records reveals that Bauman believed he could 'make 1s 6d per day' hawking spectacles.[67] However, accounts of fraud in *The Optician* and newspapers in the early 1890s reveal that traders' conduct, whether travelling or inappropriately vending from a shop, was subject to increasing attention in both large cities and more rural locations.[68] Such accounts hint at a body of traders of various backgrounds that continued to adopt more traditional techniques, such as a surviving 1884 prescription for spectacles for a J. Ellis from A. Carter of 'High Street', which includes a lens measurement in inches.[69] However, as this chapter has shown, a concerted group of opticians adopted new vision testing techniques and/or aligned their practices to other opticians across the country. This chapter has so far demonstrated that a chemist optician in Aberdeen advertised accurately prepared oculists' prescriptions, opticians in Manchester and Liverpool built vision test rooms, medical practitioners and opticians collaborated on the creation of new sight testing technologies in Newcastle and London, opticians established working partnerships with hospitals in Aberdeen and Manchester and opticians in London and others with branches in Birmingham and Glasgow wrote and printed pamphlets.

While we are unpicking the simplistic urban–rural divide, cities also faced an additional competitor to the street hawker: the automatic machine. A spectacle case in the Science Museum's Ophthalmology collection carries the image of a key and the inscription 'Automatic Sight Testing Company'. In 1889 an article in *The North-Eastern Daily Gazette* headed 'Every man his own oculist' described the installation of Mr Woolfson's automatic machines at railway stations and hospitals.[70] The article described a process in which the machine, by turning a handle, allowed the customer to cycle between twenty-two pairs of lenses of varying convexity and concavity. In 1891 Mr Woolfson set up the Automatic Sight Testing Company and became the company's managing director. How these machines worked is difficult to ascertain. Patents for other automated machines that allowed users to test their vision and purchase a pair of spectacles did appear in patent records for 'B. Green' in 1888 and 1892.[71] Yet criticism by opticians of the numbers in use was contradictory. In 1894, for example, *The Optician* declared that their use at metropolitan railway stations was decreasing.[72] In contrast, a correspondent writing to the journal in 1895 argued that, in the last twelve months, spectacle and folder sales via these machines had totalled 70,000 and the demand for them increased 'daily'.[73]

It is arguably the ideology behind the machine rather than the number in operation that was important to opticians and medical practitioners. In 1894 *The Optician* contended that 'the attempt to substitute a machine worked by a penny-in-the-slot runs counter to modern ophthalmic teaching and reduces the whole matter to the level of an absurdity'.[74] The Automatic Sight Testing Company did threaten *The Optician* with a libel suit in response to several criticisms.[75] Nevertheless, correspondence suggests that the journal's concerns were shared by medical practitioners and published in the medical press.[76] Twenty-two lens options did not provide the degree of precision necessary for accurately determining the lens that a person required, and neither would it be possible to diagnose or ascertain more complex refractive conditions, such as astigmatism. In fact, the very existence of the machines challenged new medical ideas about how to test vision and use spectacles. In contrast to the Automatic Sight Testing Company's claims, medical practitioners and opticians in late Victorian Britain argued that every man could not be his own oculist.

The certificated and qualified dispenser

To differentiate between the 'spectacle dealer' and the 'first-class optician', and to reinforce their area of expertise, a collective body of traders in *The Optician* sought certification and qualification. *The Optician*'s ambitious aims reveal a professional consciousness, not dissimilar to druggists and the establishment of the Pharmaceutical Society in the 1840s.[77] The *Optician* conceived of the journal as a pedagogical tool to upskill its readers and, like other nascent professions, presented a new knowledge of the language of science and technology to justify its claims for a specific area of expertise; it established serials – including 'the optician's library' and 'papers for beginners' – and articles in greater depth or excerpts on vision testing techniques written by opticians and medical practitioners.[78] In addition, it's editors aimed to use the journal as the mechanism for planning and shaping the establishment of an Optical Institute and certification, which received focused debate from 1893. How did medical practitioners respond? Did opticians' actions influence consumers' choices about where to obtain spectacles? While medical practitioners sought to establish a clear boundary between their practice and that of opticians in high street stores, the reality was more an entangled, triadic web that was shaped by the medical practitioner, the optician and the consumer.

Scholars have highlighted the centrality of professional periodicals, and in particular *The Lancet*, in the campaign for medical reform.[79] Parallels can therefore be drawn between the methods used by *The Lancet* and *The Optician* in their attempts to shape their respective 'professions' and create a collective body of experts. In addition, studies of medical professionalisation have highlighted the importance of local medical societies in the fostering of unity and a professional community. Such strategies helped establish an 'imagined community' between city and regional practitioners, even if this ideological community did not function or exist in practice.[80] The Victorian optician sought cohesion. Unlike S.W.F. Holloway's assessment of druggists in the 1840s, initiatives in *The Optician* engaged with provincial concerns, and did not just represent the interests of an elite group of London retailers.[81] *The Optician*'s editors and correspondents cultivated the creation of local societies and opportunities for relationships between metropole and regional businesses in a bid to create a professional community that sought to both regulate and standardise the trade. Such desire is not unique to opticians or medical practitioners and is part of a much broader concern with professionalism and being a 'professional' that emerged in a range of disciplines and among a variety of retailers during the nineteenth century.[82] The liminal position of the ophthalmologist created a unique set of circumstances. At a time when the concept of a 'profession' and the use of examinations became more important to all medical practitioners, ophthalmologists had a significant challenge in establishing their role, both outward, as public experts, and inward, within the medical profession itself, which had a mistrust for specialisation.[83] Accordingly, ophthalmologists were particularly reluctant to endorse the establishment of examining bodies that would qualify opticians to test vision when vision testing remained ill-defined in the general medical and ophthalmological curriculum. In the late 1880s, for example, optics and refraction do appear in some ophthalmology texts but their authors feel it necessary to impress the importance of the topic in their prefaces, in a bid to legitimise specialised training.[84]

Ophthalmologists acknowledged the newfound role of *The Optician* as a threat to their efforts to regulate and medicalise vision testing practice, and the 'oculist' – a pseudonym adopted by medical practitioners – quickly emerges in the journal's correspondence columns and features as part of their debates. Ophthalmologists' disapproval of opticians' methods could have very real repercussions in the early 1890s where, for example, in 1892 ophthalmic surgeons were successful in their attempts to block proposals for technical education in *The Optician*.[85] Trade journals thus also reveal the interdependent and performative nature of professions and their boundaries. As Delia Gavrus's work on twentieth-century neurology has shown, dialogue, debate and speeches published in trade journals created a sense of a shared professional community at a time of contested boundaries and interdisciplinary conflict.[86] Opticians and medical practitioners adopted pseudonyms, published their views and similarly 'performed' in *The Optician*. In 1894, for example, a Twickenham optician's correspondence demonstrated that opticians were all too aware of the significant professional implications of the publication of ophthalmologists' negative views. He argued that recent 'fatal objections' to the certification of opticians in the journal had 'sounded the trumpet of hostility' and seen 'war proclaimed'.[87] The language of war or use of metaphor adopted by opticians – a type of anger evoked by Victorian and twentieth-century medical practitioners alike – indicates the perceived 'belligerent' nature of disciplinary encroachment and the 'urgency' of stemming its influence.[88] The 'Surgeon oculist' responding to 'A Twickenham optician' presented to the public loftier aims, arguing that it was not a matter of professional jealousy and was instead about securing the welfare of the public.[89] As Ueyama has argued in relation to pharmaceutical medicines, however, it would be inaccurate to believe that medical personnel were 'primarily motivated by their professional moral code or ideals'.[90] Rather than a public service ideal that promoted the idea of 'community welfare', practitioners were hostile to the market forces controlling both the sale of pharmaceutical medicines and technological innovations.[91] In the sale of spectacles, this is also true. Ophthalmologists placed advancements in ophthalmology at the heart of this debate to mask their hostility beneath the claim that, without medical training, opticians were unable to identify disease and recognise the 'fine line' between simple and complex refractive and accommodative states of vision. Despite the editors' attempts to create a convincing and clear picture of the united 'first-class optician' in *The Optician*, the correspondence it published in the 1890s demonstrates that not all opticians agreed with the journal's aims and reforming practices. It *was* universally agreed that an optician specialised in the understanding of optical laws and the grinding of lenses, whether in their manufacture or their retail. Outside this fundamental core of an optician's practice, opticians varied on the degree to which 'the first-class optician' that *The Optician* pushed for – who fully understood optical laws, the anatomy of the eye, the use of diagnostic vision testing equipment and the prescribing of spectacles – accurately presented their level of expertise and/or current practice. A leading header 'Oculist or optician' in January 1895 argued the 'science of geometrical optics is *entirely* distinct from medical training'.[92] However, while geometrical optics might indeed have been distinct from medical training, several opticians recognised that vision testing required a blending of the two. Correspondence between 21 February and 21 March 1895 under the heading 'Opticians should be up and doing' revealed a body of opticians pushing for change and the need for appropriate training and certification to conduct the vision test.[93] Yet it also revealed that there were opticians who remained more aligned to the traditional expertise and techniques of a trader that specialised in manufacture and/or sale of optical wares; a role that they believed complemented ophthalmologists' proposal that the working relationship should involve carrying out only prescription-based work. Crucially, what side of the debate an optician would fall – 'first-rate optician' or mere 'spectacle dispenser' – was not as clear-cut as the proposed distinction. In 1896, for example, *The Optician* was particularly concerned that the latest form of opposition was from someone of 'high standing', W.A. Dixey, a London optician whose business had served successive monarchs. Dixey, writing to *The Lancet*, defended opticians' specialised status as experts in scientific knowledge of optics but argued that 'the great increase during the last thirty years in the knowledge of ocular refraction and the therapeutic use of spectacles has lifted the whole matter into the professional sphere', a professional sphere that did not include opticians but rather the medical practitioner.[94] In writing to *The Optician* to defend his stance, Dixey equally warned the journal to not be blind to the advancements in medical knowledge of the eye and vision testing.[95]

Dixey's arguments are indicative that certain opticians were moving away from an association with vision testing, in response to increasing medical involvement in the 1890s, to try and demarcate their practice from the more miscellaneous trader still functioning in the spectacle market. Yet, as might not have happened in other nascent professions, the comments of Dixey and others had limited overall impact in preventing opticians garnering a role in vision testing by carving out the occupation that would later become 'optometrist', a professional group able to test vision, prescribe lenses and refer cases of disease to the medical practitioner. This was based on both the concerted efforts of this group of opticians and the support that they received from certain ophthalmologists who recognised the value of their skillset and the demand for spectacles. Writing to *The Optician* in March 1896, for example, an oculist acknowledged the public's preference to visit an optician and conceded monopoly over routine vision testing, listing the simple cases of refraction that were the optician's 'domain' and that would ensure 'the best for their clients'.[96]

In 1897 opticians were successful in implementing early forms of certification through their newly established British Optical Association (BOA), which would later become the College of Optometrists. An additional examining body established by the Spectacle Makers' Company (SMC) – an independent livery company established in 1629 – conducted its first examination a year later, in 1898. It was initially hoped that the two examining bodies would merge, and as the SMC launched their examination a clause allowed forty-six members who had passed one of the two higher-level BOA examinations to receive the SMC diploma. However, the SMC offered a different type of examination. Opticians and ophthalmologists collaborated to produce a curriculum that limited the practical assessment to measuring or fitting lenses to a person's face based on the candidates' ability to correctly read and interpret a prescription.

By the time the first SMC examination took place, in early November 1898, *The Optician*'s view of the 'first-rate optician' and who should sit the exams had expanded. Indeed, on 5 May 1898 *The Optician* reported that the SMC committee had resolved to invite and nominate a member from the National Association of Jewellers and the Pharmaceutical Society, thus legitimising the role of the chemist and jeweller optician.[97] Four months later, in early September, *The Optician* reported on a speech delivered by Thomas Field, honourable general secretary of the National Association of Jewellers, to the Association's members in Bath.[98] Field argued that the SMC had previously ignored the 'real' optical trade and estimated that '15,000 jewellers and watchmakers' engaged in the sale of spectacles and eyeglasses. *The Optician*'s more inclusive stance towards chemists and jewellers reflects these traders' responses to the utility of certification and examination and the reality of the spectacle marketplace. Jewellers were long-standing supporters of an optician's position. In March 1895, for example, Thomas Field had addressed the National Association of Jewellers to condemn the British Medical Association's proposed bill to make it a penal offence for anyone to practice as an ophthalmic optician, who was not registered as such. Yet, while Field advised jewellers to read the textbook *Eyesight and How to Use It*, at this point his support focused more on the respective business prospects and losses if the bill was passed. He argued that the British Medical Association and the oculist-optician should not be allowed to take 'business away from them', particularly one that was 'a very remunerative part of their trade'. A variety of jewellers were reported to have supported Field's resolution, arguing that jewellers in various towns had been practising this business for decades, a claim that is further supported in jewellers' advertising in newspapers and periodicals. The chairman of the Association, for example, argued that his father and grandfather before him had sold spectacles in the north England town of Penrith for the last seventy or eighty years, and claimed the title 'ophthalmic optician'. He reflected that 'it would be a pity if they had to drop those titles, in order to increase the profits of the medical profession'.[99]

On the eve of examination in September 1898, Mr Thomas Field's assessment of jewellers' involvement in spectacle dispensing echoed more of the rhetoric of community welfare being expounded by the medical practitioner than that of the businessman. Field argued that the public 'need glasses' but that often they did not know where to go and were therefore often victims of haphazard dispensing.[100] Of Field's estimated 15,000 retail jewellers 'handling optical goods', he feared that the proportion that could pass the new examinations was 'very small'. Field was acutely aware that the SMC's scheme would result in the 'incompetent' leaving the trade and being unable to 'secure the enormous benefits' of the 'competent' trader. These benefits were not solely inward-looking – the potential business prospects of spectacle sale – but also pointed to a need for the 'public to be protected from quackery'. Field presented a two-fold argument about why jewellers should support, and also sit, the SMC's examination: 'the incentive lies not only in a safe and profitable increase in your returns, but also in the good you can do the community at large'. *The Optician* favourably supported such initiatives. In January 1899 the editors argued that the clock- and watchmaker could legitimately construct and sell visual aids because they had 'a high order of intelligence, as well as manual skill... Every skilful jeweller has in him the making of the most capable Optician.'[101]

Implicit to *The Optician*'s welcoming of jewellers and watchmakers, albeit qualified, was the need to meet the demands for certificated vision testing in the face of persistent quackery, as well as to help fill the gap that would follow the total elimination of quackery through regulation. To demonstrate its approval, the journal adopted an extended title for the 1899 volume in which these issues were published: *The Optician and Photographic Trades Review: The Organ of the Jeweller, Watchmaker, Acetylene Dealer, Photo Trader and Chemist*.[102] It is also reflective of jewellers altering their practices, as Field hoped. The surviving cash day book sales of late Victorian Carlisle optician and jeweller, John Potter Dowell, show that his practice evolved from the use of arbitrary measurements of lens strength to the making of spectacles to prescription. In the accounts between 1885 and 1894, for example, only a limited number of lens measurements were recorded – in the style 'specs no. 14', 'specs no. 2', or 'specs no. 6'. Between 1897 and 1898 there is an alteration in style: despite one mention of spectacles at 'no. 13', the customers' prescriptions were included in increasingly standard form, with the dioptric measurement.[103]

Analysis of successful candidates for the BOA and SMC examinations in *The Optician* does not provide a breakdown by trade. However, it does highlight that, in addition to a variety of trades being invited to attend examinations, the demographic of candidates passing these examinations was national in scope. The successful candidates for the BOA's March 1899 examination, for example, included opticians from Manchester, Canterbury, Sheffield, Bristol, Great Yarmouth, Oldham, Hyde, Grantown-on-Spey, Liverpool and London.[104] Similarly, for the March 1899 SMC examination, seventy-six successful candidates from 'the provinces' outnumbered fifty successful candidates from London.[105] Not everyone passed the examinations, and such findings point to a variation in levels of expertise. Nevertheless, in December 1899, for example, *The Optician* reported that they were 'very pleased' that 'so many' as sixty-one candidates had been successful out of the ninety-six who sat the SMC's examination in November.[106]

How did medical personnel respond to a diverse body of traders, united in their title of optician, sitting the BOA and SMC examinations? George Lindsay Johnson, a registrar at the Royal Westminster Ophthalmic Hospital and later ophthalmologist at the Royal Eye Hospital, was an influential figure in devising optician's training and examination and a vocal advocate for opticians in the 1890s. Studying his practice and both his private and public opinion, from his collaborative devising of the SMC examination syllabus in 1898 up until 1904 – when the SMC curriculum altered to include vision testing – reveals the extent and limitations of medical practitioners' arguments. In 1898, ophthalmologists held considerable influence. Johnson had attested to opticians' 'large experience' and involvement in vision testing but conceded that this view was not 'shared by the medical profession in general'.[107] Initially, he therefore advised that opticians consider the eye 'merely as an optical instrument' in the SMC's examinations.

Johnson's assessment was correct, and the SMC's first syllabus was approved by both *The Lancet* and the *British Medical Journal*.[108] However, it was not long before the threat of certificated opticians became a matter of debate in the *British Medical Journal*.

Traders, naturally, advertised that they had passed the examinations set up by the BOA and SMC in the late 1890s. Between 1899 and 1900, for example, C. Wallace described himself as a 'qualified Optician & Watchmaker' and a 'Member of the British Optician Association' in *The Isle of Man Times and General Advertiser*.[109] Correspondents in the *British Medical Journal* presented advertising opticians as lying or posing as medical men when they passed the SMC or BOA examinations – particularly if they went on to use the letters 'FSMC' or 'BOA' after their name – because this could suggest to an ignorant public that they had received a form of medical training.[110] Medical practitioners' criticisms targeted the trade and commercial nature of an optician's practice. Focus on opticians' advertising, on one hand, exposed the tensions between medical professional practice and, as Ueyama has identified in broader medical practice, the 'tug of war' over profits that ensued.[111] On the other hand, it also demonstrated the threat of certification in helping to promote an optician's public standing. The SMC initially conducted itself carefully in response to these early claims. It developed an Advertisement Sub-Committee to monitor opticians' conduct and made candidates sign a legal agreement not to use any form of drug or to advertise the diploma 'in any way which would lead the public to infer it conveys medical qualifications' and breach the Medical Acts.[112]

In March 1899 the influence of medical practitioners' opinions on the SMC and control over opticians' conduct was waning. The master of the SMC, while in favour of proper conduct, questioned the legitimacy of medical intervention, arguing that the Medical Acts could not prohibit opticians from learning their own business.[113] Such a stance only intensified medical retaliation. In May 1899 *The Optician* struggled to interpret the decision of medical practitioners to send a memorial to the General Medical Council on the subject of 'the issue to persons not medically qualified, of certificates of proficiency in departments of medical practice'.[114] Here, practitioners grouped opticians' continued practice in vision testing with other nascent professions, such as midwifery. As outlined in Roger Cooter's analysis of the relationship between nineteenth-century bonesetters and early orthopaedists, medical practitioners did not always seek to deny or dismiss alternative practices but could legitimise them as a form of medical practice in order to appropriate and exploit them. Ophthalmologists, too, were attempting to take over a role that garnered 'popular and professional' respect' by claiming that it needed to be under the control of the medically trained.[115] Correspondence in the *British Medical Journal*, for example, sought to undermine opticians' expertise by trying to expose instances where a 'qualified' optician had made an error or had failed to provide a person with the correct prescription. In July 1902, for example, a correspondent argued that an optician had prescribed glasses that were 'altogether wrong' and that the diplomas were 'responsible for the creation of a set of charlatans who, under the *aegis* of such diploma, mislead and impose on the public'.[116] The content and intensity of medical practitioners' responses demonstrates their hostility to the broader growth of medical capitalism and the commercialisation of medical products.[117] It also hints at opticians' initial success in challenging the legitimacy of medical authority over matters of vision testing. In 1903 Johnson's advice to the SMC changed. He wrote to the *British Medical Journal* stating that he was stepping down as an examiner for the SMC until medical practitioners acknowledged that the optician's trade 'is not, and indeed cannot be, confined within the present limits of the SMC examination'.[118]

In addition to his public, vocal stance, Johnson's private practice indicates how positive working relationships could be established with opticians at the turn of the twentieth century. A set of letters exchanged between Johnson and London-based optician James Aitchison between 1898 and 1900 reveal an equal partnership of cooperation.[119] The letters demonstrate that Aitchison could prescribe glasses and send them to Johnson to confirm, and Johnson could ask Aitchison to try to get a better prescription for the patient. In a letter from 7 June 1898, for example, Johnson discussed a patient's eyesight and stated 'I cannot get her vision better than 20/40 to 20/5 R & L... I wish you to see if you can improve her sight'. The trade directory entries for Aitchison are also revealing and denote how he removes any potential association with medicine from his firm's name; in 1899 he dropped the title of 'oculist' and was listed only as an optician.[120] In 1904, on the basis of his partnership with Johnson and other London medical personnel Aitchison became vocal in the 'Testing: the eyesight of a nation' debate that emerged in the *British Medical Journal* as the SMC was devising its new syllabus and proposed to include vision testing. He argued that both opticians and ophthalmologists were adjusting to recent changes to vision testing and that opticians needed to increase their studies of optics and refraction while ophthalmic surgeons increased their studies of eye disease, so that a favourable working partnership could be established, one that aimed to promote the advancement of knowledge in, and collaboration between, both disciplines.[121] Aitchison's working relationship with Johnson suggests his proposal was a practical working approach that did not involve the physical separation of their two practices but instead acknowledged each body's specialised skillset and engaged in cooperative communication.

Aitchison and Johnson may have been atypical. There were opticians that sought to advertise deeper medical associations and there were opticians that did not dispense spectacles or conduct a vision test. Equally, there were practitioners that disregarded opticians' skillset and practitioners that openly admitted that an optician was more qualified, and had a greater understanding of refraction, to test people's vision. The outcome of these debates was thus one of grudging acceptance when the SMC was successful in including vision testing in its practical examination in 1904. While in response to Aitchison's correspondence in the *British Medical Journal* a medical practitioner argued that 'of all medical impositions on the public that of the *shopkeeper* deciding upon glasses is probably the most pernicious', such emphasis on an optician's trading background did little to curb the dominant practice of dispensing spectacles on the high street.[122] Doctors, like the broader market for pharmaceuticals and medical commodities, could not remain indifferent to the reality of spectacle retail.[123] In Aitchison's defence another practitioner, F. Dugon, argued that individuals chose not to attend an ophthalmic surgeon because, on one hand, the cost was higher but, on the other, they simply 'did not wish to go to hospital for errors of refraction'. In response to the public's preference, Dugon argued that his collaboration with Aitchison was a positive one because 'refraction is corrected perfectly by the competent optician'.[124] Indeed, he suggested several practitioners had adopted this practice.

With the preference of the public in mind, opticians were uniquely positioned in the 1890s and early 1900s to adopt certification to consolidate their technical expertise in vision testing and effectively challenge the legitimacy of the ophthalmologists' position. The retail market challenged medical control over the commodification of spectacles as a prosthesis and positioned spectacle users as active agents.[125] Far from patients or consumers being passive, ophthalmologists' and opticians' comments highlight that patients', or customers', choices shaped the place in which the vision test, and new technologies, would be administered. Ueyama's study of late Victorian London demonstrated that medical practitioners and consumers were equal partners in the medical marketplace.[126] In her study of nineteenth-century professionalism Penelope Corfield similarly puts forward the idea that consumers were not oppressed, or submissive, and had their own autonomy, an autonomy that could evade or offer competing interpretations.[127] On the advent of certification spectacle users did not seek the optician's shop just as an alternative to the hospital. Like other consumers in the market, they had a choice. As Anderson has shown in her study of artificial limbs, the proliferation of newspaper advertisements for prosthetics reveal that this was a commercial choice.[128] A prospective customer's decision was motivated by several factors, not just medical practitioners' and opticians' preoccupation with an appropriate vision test and accurate lenses. Style, design and popular trends were important. Irvine Loudon, for example, has argued that public fashions controlled the early nineteenth-century market for drugs.[129] Similarly, just as it encouraged the consultation of an optician in lieu of an optician, price also continued to shape the fringes of the spectacle market. An article on 'cheap spectacles' in *The Optician* argued that the public 'created the imposter' and encouraged the practices of the 'quack' because of their preference for cheap and inappropriately sold products.[130] It is impossible to know how many sought the opinion of a qualified optician and/or a medical practitioner. But, in 1898 correspondence in *The Optician* exposed the perceived scale. The letter writer claimed that at least 90 per cent of the public did not follow reformed spectacle dispensing practices and argued:

['the ordinary individual'] at the present time proceeds to buy a pair of glasses much in the same manner as he would do a pound of tea; and he is seen more concerned as to the quality of the tea than the suitability of the glasses he may select. His chief endeavour is to buy glasses as cheaply as he possibly can, while he will buy good tea and recognise the necessity of paying a decent price for it.[131]

Would 'the ordinary individual' have been unaware of new medical advancements in the eye and the increasing role of medical knowledge in the dispensing of spectacles? Probably not. Later that year Birmingham optician H.L. Taylor – in his three-part article 'Fraud in the optical trade and how to combat it' – argued that the biggest threat to opticians' practice was 'quacks' posing as medical practitioners and adopting the use of testing equipment; 'armed generally with a cheap sliding optometer it is easy to imagine the harm such a person can do'.[132] A week later, Taylor suggested that the 'general form of the quack nuisance in the larger towns may be styled "The Professor"'. The use of sight testing technologies was also pervasive in fringe advertising. Taylor, for example, draws a bead on one hawker who 'spreads his fame by means of photographs representing him standing in dramatic attitude by a trial case, the contents most ostentatiously displayed'.[133] Similar to other quack or fringe medical practices, retailers or practitioners fashioned their 'new garb' in response to medical and cultural change.[134]

If we compare the spectacle hawker of the 1830s, impressing upon the public his knowledge of the laws of optics and basic anatomy of the eye, to the hawker of the 1890s, a specialist in eyesight and with 'the most audacious fabrications as to his qualifications', it is clear that new medical ideas and the language of expertise filtered into the fringes of spectacle dispensing practices. The increasing professionalisation of opticians' practice, ophthalmologists' involvement in vision testing and the advent of certification had far-reaching effects. Consumers were not foolishly enticed by the unqualified. Both opticians and practitioners highlight spectacle users' choices and their ability to shrewdly determine best price. As argued by Patrick Wallis in his study of the early modern medical economy, it was difficult to evaluate the quality of drugs, even for a skilled consumer who exercised discernment.[135] It would have been equally challenging for a late Victorian consumer to measure the efficacy of a pair of spectacle lenses. While consumers could seek reassurance in opticians' qualifications and/or their relationship with the medical profession, they could also exercise their own judgement. The prospective spectacle wearer perceived the obtaining of spectacles as a transaction in which the customer sought to obtain the best price for goods of a certain quality, and not necessarily a service; the latter was determined by the quality of the vision test and the knowledge or expertise of the dispenser. Such a judgement blurred the boundaries between an optician's practice and a medical practitioner's. Rather than fixed battlelines, it embodied a dynamic relationship involving a variety of historical actors determining the relative success and ultimate limits of medical efforts.

Conclusion

Birmingham optician H.L. Taylor prepared *The Practical Optician's Guide: An Elementary Course for Opticians* as a textbook for opticians studying for the anticipated examinations. Writing the preface in December 1896, he argued:

> My object has been to fit in necessary information between two limits. The one, a certain vague rule of thumb method practised still by probably the majority of opticians; the other, the minimum knowledge required by the careful operator.
>
> Several matters are therefore discussed at greater length than under other circumstances would be advisable, notably the comparison between the dioptric system and the methods of measuring lenses by the unskilled opticians.[136]

Taylor's observations, while only an estimate, propose that prior to the advent of examinations the 'majority' of opticians adopted a haphazard method. By 1904, and on the advent of vision testing being introduced into the SMC examination, the council of the BOA estimated that 600 opticians had been certificated. The demographic of successful candidates was national in scope. Certification, as predicted, promoted opticians' practice, and legitimised their role in vision testing and spectacle dispensing. It was, however, a watershed moment and not a victory. It initiated debates about appropriate conduct between opticians and medical practitioners that would continue for half a century, until the Opticians Act of 1958, an act that saw a body of opticians succeed in establishing their role as optometrists and gain a professional register. In 1904, in addition to the 600 certificated, the BOA estimated that a further 20,000 non-certified traders continued to dispense spectacles. Indeed, in the same year, a correspondent using the pseudonym 'BSc., MD' in the *British Medical Journal*'s 'Testing the eyesight of a nation' debate exposed a case of the inch measurement system being used, a form of measurement that Taylor had deemed 'unskilled'. The case involved the reporting of an optician by a lady 'at a fashionable seaside town on the South Coast'. When confronted by 'BSc., MD' the optician 'insisted that the inch-system was in use', and that the dioptric system was 'wrong'.[137]

The account in the *British Medical Journal* further reinforces the argument put forward that a prospective patient and/or consumer was neither uneducated nor unaware of advancements in vision testing and spectacle dispensing. The lady, in this case, evidently knew more than the optician and reported him. However, that did not mean that all spectacle users then sought the advice of certified opticians or medical personnel. It is important to consider the logic of consumers' choices when the number of qualified retailers would have been small. A late Victorian consumer wishing to purchase a pair of spectacles had multiple options. They could have consulted an ophthalmologist in a hospital, an optician or a variety of other traders on the high street, a travelling hawker on the street or in temporary premises or have operated an automatic machine at a railway station or busy thoroughfare. They could have a thorough eye examination and vision test that involved a variety of new sight-testing technologies or have been able to exercise their own judgement and try different lenses to choose a pair that they believed suited them best. The dispenser could have attracted a potential customer either by advertising, or proving, their expertise in vision testing, the quality of their frames and lenses, or their competitive price. The consumer's decision would have been influenced not just by their perceived declining vision and need for lenses. Spectacle users were navigating a complicated market where the functionality of the frame was not necessarily a primary interest; price, design and style were equally important. Spectacles were a commodified assistive device, how they looked and how much they cost drove consumer demand and consumer choices. Opticians were better placed than medical practitioners to respond to market demands and a working partnership that enabled medical personnel to engage with this market is something that the practitioners had to grudgingly accept. The range of meanings that spectacles embodied – as Chapter Five will demonstrate – influenced the sites of spectacle sale. Spectacles were not a solely functional, medical device and their users were reluctant to see them as such. Just as today, spectacle sale thrived on the high street.

Notes

1 'The Ophthalmic Institution and what I saw there [by a patient]', *Aberdeen Weekly Journal* (14 December 1894), reprinted 19 December 1894.
2 Claire L. Jones, *The Medical Trade Catalogue in Britain, 1870–1914* (London: Pickering and Chatto, 2013).
3 Takahiro Ueyama, *Health in the Marketplace: Professionalism, Therapeutic Desires, and Medical Commodification in Late-Victorian London* (California: SPSS, 2010).
4 Julie Anderson, 'Separating the surgical and commercial: space, prosthetics and the First World War', in Claire L. Jones (ed.), *Rethinking Modern Prostheses in Anglo-American Commodity Cultures 1820–1939* (Manchester: Manchester University Press, 2017), pp. 158–178; Caroline Lieffers, 'Itinerant manipulators and public benefactors: artificial limb patents, medical professionalism and the moral economy in antebellum America', in Jones (ed.), *Rethinking Modern Prostheses*, pp. 137–157.
5 Jones (ed.), *Rethinking Modern Prostheses*.
6 For the eighteenth century see, for example, Liliane Hilaire-Pérez and Christelle Rabier, 'Self-machinery? Steel trusses and the management of ruptures in eighteenth-century Europe', *Technology and Culture*, 54:3 (2013), 460–502 (p. 490).
7 Patrick Wallis, 'Introduction: the growth of the early modern medical economy', *Journal of Social History*, 49:3 (2016), 477–483 (p. 477).

8 S.W.F. Holloway, 'The orthodox fringe: the origins of the Pharmaceutical Society of Great Britain' in W.F. Bynum and Roy Porter (eds) *Medical Fringe & Medical Orthodoxy 1750–1850* (London: Croom Helm, 1987), pp. 129–157, 137.

9 Anderson, 'Separating the surgical and the commercial', p. 172.

10 Holloway, 'The orthodox fringe'; see also Irvine Loudon's and Roger Cooter's chapters in Bynum and Porter (eds), *Medical Fringe & Medical Orthodoxy*.

11 'Presentation to Sir Anderson Critchett and Mr Malcolm Morris', *British Medical Journal* (28 November 1903), p. 1432.

12 'Obituary: Sir Anderson Critchett', *British Medical Journal* (14 February 1925), p. 338.

13 Jones, *The Medical Trade Catalogue*, pp. 21–22, 24.

14 Percy Dunn, 'A model eye', *British Medical Journal* (10 January 1903), p. 88; Science Museum's Ophthalmology collection, object number A681267.

15 Christopher Jeaffreson comments on his improved instrument in response to another optometer discussed in the BMJ: 'Clinical memoranda – Dr Laidlaw Purves's new optometer', *British Medical Journal* (30 January 1875), p. 140.

16 Joel Soelberg Wells, *On Long, Short and Weak Sight and their Treatment by the Scientific Use of Spectacles* (London: J.A. Churchill, 1862), pp. 111–112.

17 Devon Heritage Centre, 1299F/0/HA/19, West of England Eye Infirmary, cashbook of subscriptions, 1885–1895.

18 Jones, *The Medical Trade Catalogue*, p. 152.

19 James Weiss & Son, *A Catalogue of Surgical Instruments, Apparatus, Appliances etc* (1863); *Illustrated Catalogue and Price List of Surgical Instruments and Appliances* (1889); *A Catalogue of Ophthalmic Instruments and Appliances* (1898).

20 Down Bros, *A Catalogue of Surgical Instruments and Appliances* (1885; 1887; 1889; 1892; 1894; 1896; 1897; 1900); *A Catalogue of Surgical Instruments and Appliances, also of Aseptic Hospital Furniture* (1901).

21 Comparisons can be made here to the artificial limb maker: see Anderson, 'Separating the surgical and the commercial', pp. 158–178.

22 Jones, *The Medical Trade Catalogue*, pp. 9, 48.

23 Reynolds & Branson, *Handy Guide to Surgical Instruments and Medical Appliances, Physical and Physiological Apparatus, Microscopes, Spectacles, Artificial Eyes* (1887), pp. 237–241.

24 Brady & Martin, *Illustrated Catalogue of the Instruments, Appliances, and Fittings Used in Surgery & Medicine* (1897), p. 225.
25 'Lindsey and Sons', *Jackson's Oxford Journal* (7 September 1889).
26 Science Museum's Ophthalmology collection, object number A49646.
27 *History and General Directory of the Borough of Sheffield with Rotherham, Chesterfield, and all The Villages and Hamlets* (1833), p. 149; *General Directory of Sheffield* (1849), p. 177; *Gazetteer and General Directory of Sheffield* (1852), p. 181; *General and Commercial Directory and Topography of the Borough of Sheffield* (1862), pp. 47, 149; *White's General and Commercial Directory of Sheffield* (1879), pp. 809, 397; *Kelly's Directory of West Riding of Yorkshire* (1881), p. 1950; *White's Directory of Sheffield & Rotherham* (1901), p. 5930.
28 *Aberdeen Weekly Journal*, appears weekly between 1 January 1895 and 27 December 1895.
29 Science Museum's Ophthalmology collection, object number A682408. This case is undated; however, L.J Troulan appear as an optician at 11 Higher Union Street in Kelly's *Directory of Devon* (1889), p. 588.
30 *Manchester Times*, weekly between 7 June 1890 and 22 August 1890.
31 *Liverpool Mercury* (18 April 1895).
32 Mary Carpenter, *Health, Medicine and Society in Victorian England* (Santa Barbara, CA: Praeger, 2009), pp. 2, 4–5, 12; Stanley Joel Reiser, *Medicine and the Reign of Technology* (Cambridge: Cambridge University Press, 1978).
33 Robert Brudenell Carter, *Eyesight: Good and Bad: A Treatise on the Exercise and Preservation of Vision*, 2nd edn (London: Macmillan, 1880), p. 59.
34 See, for example, N. David Richards, 'Dentistry in England in the 1840s: the first indications of a movement towards professionalisation', *Medical History*, 12 (1968), 137–152; Eric G. Forbes, 'The professionalisation of dentistry in the United Kingdom', *Medical History*, 29 (1985), 169–181; S.W.F. Holloway, 'Producing experts, constructing expertise: the School of Pharmacy of the Pharmaceutical Society of Great Britain, 1842–1896', in Vivian Nutton and Roy Porter (eds), *The History of Medical Education in Britain* (Amsterdam: Rodopi, 1995), pp. 116–140; Daniel J. Malleck, 'Professionalism and the boundaries of control: pharmacists, physicians and dangerous substances in Canada, 1840–1908', *Medical History*, 48:2 (2004), 175–198; Geoffrey Stuart Taylor and Malcolm Nicolson, 'The emergence of orthodontics as a speciality in Britain: the role of the British Society for the Study of Orthodontics', *Medical History*, 51 (2007), 379–398; Christine

E. Hallett, 'Nursing, 1830–1920: forging a profession', in Anne Borsay and Billie Hunter (eds), *Nursing and Midwifery in Britain since 1700* (Basingstoke: Palgrave Macmillan, 2012), pp. 46–73.

35 See also Richards, 'Dentistry in England', 237–8; S.E.D. Shortt, 'Physicians, science and status: issues in the professionalisation of Anglo-American medicine in the nineteenth century', *Medical History*, 27 (1983), 51–68 (pp. 51–2).

36 Andrew Abbott, *The System of Professions: An Essay on the Division of Expert Labour* (Chicago: The University of Chicago Press, 1988), pp. 1–9.

37 George Weisz, 'The emergence of medical specialisation in the nineteenth century', *Bulletin for the History of Medicine*, 77 (2003), 536–575; George Weisz, *Divide and Conquer: A Comparative History of Medical Specialisation* (New York: Oxford University Press, 2006); George Rosen, *The Specialisation of Medicine with Particular Reference to Ophthalmology* (New York: Froben Press, 1944); Stephen Casper, *The Neurologists: A History of a Medical Specialty in Modern Britain, c.1789–2000* (Manchester: Manchester University Press, 2016), pp. 60–64.

38 Jones, *The Medical Trade Catalogue*, p. 9.

39 Ibid., p. 6.

40 Jones, *Rethinking Modern Prostheses*, p. 2.

41 'Introducing *The Optician*', *The Optician* (2 April 1891), p. 2.

42 'Correspondence – Fair Play', *The Optician* (25 June 1891), p. 205.

43 See, for example, *The Bristol Mercury*, weekly between 16 June 1877 and 21 July 1877.

44 *The Critic* (15 March 1852), p. 163; and see also the *Examiner*, weekly between 4 April 1863 and 9 May 1863; *Northern Echo*, regularly between 27 September 1895 and 14 October 1895.

45 Ueyama, *Health in the Marketplace*, p. 5.

46 Henry Laurance, *The Eye in Health and Disease: With Hints on the Choice and Use of Spectacles*, 3rd edn (London: Love Brothers, 1888).

47 Jewellers and opticians associated with Laurance: *Aberdeen Weekly Journal* (3 July 1878); *Isle of Wight Observer*, weekly between 10 April 1880 and 16 October 1880; *The Star* (19 July 1881); *The North-Eastern Daily Gazette* (3 July 1895); cutlers and opticians associated with Laurance: *Trewman's Exeter Flying Post or Plymouth and Cornish Advertiser*, weekly between 12 January 1895 and 28 December 1895; *The York Herald*, between 3 August 1878 and 11 November 1878; chemist opticians associated with Laurance: *The*

Hampshire Advertiser, weekly between 8 May 1880 and 12 March 1881; The Wrexham Advertiser, and North Wales News (26 March 1887), (24 December 1887); Aberdeen Weekly Journal, regularly between 1 January 1890 and 17 December 1890; and weekly between 2 January 1895 and 13 November 1895.
48 'The ills that flesh is heir to', Belfast News-Letter (11 November 1863).
49 'Scraps', Manchester Times (16 December 1882). 'Factae', The Royal Cornwall Gazette Falmouth Packet, Cornish Weekly News, & General Advertiser (5 January 1883).
50 George Black, Eyesight and How to Care for it (London: Ward, Lock, 1888), p. 10.
51 Ueyama, Health in the Marketplace, p. 85.
52 John Browning, Our Eyes and How to Preserve Them from Infancy to Old Age. With Special Information About Spectacles, 7th edn (London: Chatto & Windus, 1887).
53 'Reviews and notices', British Medical Journal (9 October 1886), p. 682.
54 J. Gray Keith, Facts Relating to Spectacles: How, When, Where and Why to Wear Them, with Rules for the Preservation of the Sight, And Cautions as to how thousands Ruin their Sight (Glasgow: David Bryce & Son, 1889).
55 Ibid., p. 8.
56 Anne Digby, The Evolution of British General Practice, 1850–1948 (Oxford: Oxford University Press, 1999), pp. 35, 99–100, 230.
57 Holloway, 'The orthodox fringe', p. 137.
58 'Answers to correspondents', Bow Bells: A Magazine of General Literature and Art for Family Reading (10 January 1890), p. 2.
59 Sally Shuttleworth and Geoffrey Cantor (eds), Science Serialized: Representations of the Sciences in Nineteenth-Century Periodicals (London: The MIT Press, 2004), pp. 2, 4; Gregory Tate, 'Researching science and periodicals: satire and scientific jargon in Punch', in Alexis Easley, Andrew King and John Morton (eds), Researching the Nineteenth-Century Periodical Press: Case Studies (London: Routledge, 2018), p. 161.
60 For early examples, see 'Spectacles', The York Herald and General Advertiser (21 October 1843) and Richard Middlemore, Introductory lecture on the Anatomy, Physiology, and Diseases of the Eye, Delivered at the Birmingham Royal School of Medicine and Surgery, 4 October 1839 (London: S. Longman, Orme, Brown, Green and Longmans, 1839), p. 17.

61 'Chats with housekeepers – take care of your eyes', *The Newcastle Weekly Courant* (5 July 1890).
62 Ueyama, *Health in the Marketplace*. For overview see pp. 5, 70, 74.
63 'Correspondence: a serious matter for the public', *The Optician* (18 August 1892), p. 374.
64 *Slater's Directory of Manchester & Salford* (1879), p. 10 [digital p. 68]; ibid. (1883), p. 10; ibid. (1895), p. 16.
65 University of Nottingham Archives, Manuscript and Special Collections, Wr/D/28, Diaries of Edward M. Wrench, 1856–1912, 1883, p. 93 and Wr/D/28/4, Notes. Originally found enclosed in Wr D 28, inside front cover.
66 A similar argument has been made by Patrick Wallis in relation to pharmaceutical shop design: 'Consumption, retailing and medicine in early modern London', *Economic History Review*, 61:1 (2008), 26–53 (p. 44).
67 Cambridgeshire Archives, K1350/1/892, Wilhelm Bauman aged 51, moulder. Requests stock of spectacles for hawking. Passage paid to return to Germany, 1884–1886.
68 See, for example, 'Wholesale swindling in Fife. A bogus oculist professor', *The Dundee Courier & Argus* (4 March 1893); a report on numerous frauds in Manchester, *The Optician* (8 December 1892), p. 196; and fraud in Hastings, *The Optician* (12 April 1894), p. 28.
69 Devon Heritage Centre: 1695M/FP/10, prescription for spectacles to J. Ellis from A. Carter, 1884.
70 'Every man his own oculist', *The North-Eastern Daily Gazette* (15 July 1889).
71 B. Green, 1888, patent number 16,090 in *Patents for Inventions: Abridgements of Specifications, Class 97 Philosophical Instruments* (London: Darling & Son, 1896), p. 163; B. Green, 1892, patent number 3967 in *Patents for Inventions: Abridgements of Specifications, Class 97 Philosophical Instruments* (London: Darling & Son, 1898), p. 131.
72 'Automatic sight testing', *The Optician* (20 December 1894), p. 198.
73 'Correspondence – automatic sight testing machines', *The Optician* (31 January 1895), p. 282.
74 'Penny-in-the-slot opticians', *The Optician* (6 April 1894), p. 13.
75 'Automatic sight testing', *The Optician* (20 December 1894), p. 198.
76 'Penny-in-the-slot opticians', *The Optician* (6 April 1894), p. 13.
77 Holloway, 'The orthodox fringe', pp. 130, 133–135.
78 See, for example, ibid., pp. 140–2; Jaipreet Virdi, 'From the hands of quacks: aural surgery, deafness, and the making of a surgical specialty

in 19th century London' (2014), unpublished thesis, University of Toronto, pp. 32–33; Malleck, 'Professionalism and the boundaries of control'; Taylor and Nicolson, 'The emergence of orthodontics'.
79 Ian A. Burney, 'Medicine in the age of reform', in Arthur Burns and Joanna Innes (eds), *Rethinking the Age of Reform: Britain 1780–1850* (Cambridge: Cambridge University Press, 2003), pp. 163–164; Michael Brown, 'Medicine, reform and the "end" of charity in early nineteenth-century England', *English Historical Review*, 124:511 (2009), 1353–1388 (pp. 1379–1380).
80 Michael Brown, *Performing Medicine: Medical Culture and Identity in Provincial England, c.1760–1850* (Manchester: Manchester University Press, 2010), pp. 194–195.
81 Holloway, 'The orthodox fringe', p. 153.
82 See, for example, Lieffers, 'Itinerant manipulators and public benefactors', pp. 137–157; Philippa Levine's discussion of professionalisation in relation to historical practice, *The Amateur and the Professional: Antiquarians, Historians and Archaeologists in Victorian England, 1838–1886* (Cambridge: Cambridge University Press, 2002); and for discussion of retailers and reputability: John Benson, 'Drink, death and bankruptcy: retailing and respectability in late Victorian and Edwardian England', *Midland History*, 32:1 (2007), 128–140; John Stobart, Andrew Hann and Victoria Morgan, *Spaces of Consumption: Leisure and Shopping in the English Town, c.1680–1830* (London: Routledge, 2007).
83 Keir Waddington, 'Mayhem and medical students: image, conduct, and control in the Victorian and Edwardian London teaching hospital', *Social History of Medicine*, 15:1 (2002), 45–64 (pp. 50, 63); Weisz, 'Emergence of medical specialisation'; Weisz, *Divide and Conquer*; Rosen, *The Specialisation of Medicine*; Casper, *The Neurologists*, pp. 60–64. See also for more general studies: H. Perkin, *The Rise of Professional Society: England since 1880* (London: Routledge, 2002); J.P.C. Roach, *Public Examinations in England, 1850–1900* (Cambridge: Cambridge University Press, 1971).
84 George A. Berry [Royal College of Physicians of Edinburgh], *Diseases of the Eye: A Practical Treatise for Students of Ophthalmology*, 1st edn (Edinburgh and London: Young J. Pentland, 1889); Gustavus Hartridge, FRCS [assistant surgeon to the Royal Westminster Hospital], *The Refraction of the Eye: A Manual for Students* (London: J & A Churchill, 1884), pp. iii–iv.
85 'The Spectacle Makers' Company and technical training', *The Optician* (10 March 1892), pp. 385–386.

86 Delia Gavrus, 'Men of dreams and men of action: neurologists, neurosurgeons and the performance of professional identity, 1920–1950', *Bulletin of the History of Medicine*, 85 (2011), 57–92 (p. 61).
87 'The voice of the trade', *The Optician* (18 January 1894), p. 161.
88 Gavrus, 'Men of dreams', p. 68; Irvine Loudon, 'The vile race of quacks with which this country is infested', in Bynum and Porter (eds), *Medical Fringe & Medical Orthodoxy*, p. 106.
89 'The voice of the trade', *The Optician* (25 January 1894), p. 171.
90 Ueyama, *Health in the Marketplace*, p. 57.
91 See also, the earlier example of druggists in Loudon, 'The vile race of quacks' and Holloway, 'The orthodox fringe'.
92 'Optical specialists and medical quacks', *The Optician* (7 February 1895), p. 294.
93 'The voice of the trade: opticians should be up and doing', *The Optician* (21 February 1895), p. 323.
94 'A house divided against itself?', *The Optician* (4 June 1896), p. 164.
95 'Opticians and oculists', *The Optician* (11 June 1896), p. 176.
96 'Can opticians be trusted to correct defects of sight? By an oculist', *The Optician* (26 March 1896), p. 14.
97 'The Spectacle Makers' Company committee', *The Optician* (5 May 1898), p. 254.
98 'The Worshipful Company of Spectacle Makers and the certification of opticians, delivered by Mr Thos Field', *The Optician* (1 September 1898), p. 756.
99 'Jeweller opticians', *The Optician* (21 March 1895), pp. 381–382.
100 'The Worshipful Company of Spectacle Makers and the certification of opticians, delivered by Mr Thos Field', *The Optician* (1 September 1898), p. 756.
101 'Optics and jewellery – notes on the science of decorative art', *The Optician* (5 January 1899), p. 530.
102 *The Optician and Photographic Trades Review*, vol. XVI (London: The Gutenberg Press Ltd, 1899).
103 Carlisle Archive Centre, DB9/1–7, John Potter Dowell, cash and day book (sales), 1885–1894.
104 'The British Optical Examination – results', *The Optician* (9 March 1899), p. 794.
105 'Some statistics concerning the examination scheme of the Worshipful Company of Spectacle Makers', *The Optician* (30 March 1899), p. 37.
106 'The examination', *The Optician* (15 December 1899), p. 484.

107 For an overview of the debates, see 'The voice of the trade', *The Optician* (23 April 1891), p. 62 and 'The voice of the trade', *The Optician* (9 July 1891), p. 238.
108 'The revised syllabus', *The Optician* (2 February 1899), p. 644.
109 *The Isle of Man Times and General Advertiser*, between 16 September 1899 and 29 December 1900.
110 'The Yorkshire Optical Society', *British Medical Journal* (24 June 1899), p. 1547.
111 Ueyama, *Health in the Marketplace*, p. 22.
112 'The diplomas for opticians – letter from the master of the Spectacle Makers' Company', *The Optician* (16 March 1899), p. 842, and 'The committee meeting of the Spectacle Makers' Company on the recent examination' (4 May 1899), p. 294.
113 'The diplomas for opticians – letter from the master of the Spectacle Makers' Company', *The Optician* (16 March 1899), p. 842.
114 'A plea for peace', *The Optician* (11 May 1899), pp. 323–324.
115 Roger Cooter, 'Bones of contention? Orthodox medicine and the mystery of the bone-setter's craft' in Bynum and Porter (eds), *Medical Fringe and Medical Orthodoxy*, pp. 159, 166–167.
116 *British Medical Journal* (12 July 1902), p. 156.
117 A similar exposé of pharmaceuticals is discussed in Ueyama, *Health in the Marketplace*, p. 57.
118 'Sight testing by opticians', *British Medical Journal* (11 July 1903), 109–110.
119 Boots Archives, DA15/43, collection of letters from Lindsay Johnson to James Aitchison, 1898–1900.
120 *Post Office London Trades Directory* (1895) [Part 3: Commercial & Professional Trade Directory], p. 771; ibid. (1899) [Part 3: Commercial & Professional Trade Directory], p. 853.
121 'Correspondence: testing the eyesight of a nation', *British Medical Journal* (14 May 1904), p. 1170.
122 'Correspondence: testing the eyesight of a nation', *British Medical Journal* (28 May 1904), p. 1288.
123 Ueyama, *Health in the Marketplace*, p. 6.
124 'Correspondence: testing the eyesight of a nation', *British Medical Journal* (4 June 1904), p. 1345.
125 Jones, *Rethinking Modern Prostheses*, p. 2.
126 Ueyama, *Health in the Marketplace*, p. 22.
127 Penelope Corfield in Ueyama, *Health in the Marketplace*, p. 280.
128 Anderson, 'Separating the surgical and the commercial', pp. 158–178.
129 Loudon, 'The vile race of quacks', p. 122.

130 'Cheap spectacles', *The Optician* (20 August 1896), p. 310.
131 'Correspondence – the certification of opticians', *The Optician* (7 April 1898), p. 144.
132 'Fraud in the optical trade and how to combat it – I. The pedlar and hawker', *The Optician* (8 December 1898), p. 436.
133 'Fraud in the optical trade and how to combat it – II. The professor and his ilk', *The Optician* (15 December 1898), p. 464.
134 Ueyama, *Health in the Marketplace*, p. 5.
135 Wallis, 'Consumption, retailing and medicine', p. 26.
136 Harry L. Taylor, *The Practical Opticians Guide: An Elementary Course for Opticians* (Birmingham: J. & H. Taylor, 1897), preface.
137 'Correspondence: Testing the eyesight of a nation', *The British Medical Journal* (4 June 1904), p. 1345.

5

Fashioning the eye and seeing, 1830–1904

The collections of spectacles and eyeglasses in the Science Museum's off-site stores offer an encounter with a variety of Victorian visual aid designs: the decorative, 'showy' frames, the utilitarian, workaday frames and the discreet, nondescript styles that were marketed as 'invisible'. The focus of this book hitherto has been to explore spectacles as a functional, assistive device to expose why and how a variety of historical actors held off ophthalmologists' medicalising aims. In this chapter we turn to the Victorians who wore and commented upon the visual aid to demonstrate that spectacles embodied more than their functional purpose – an argument that has been peppered throughout the preceding four chapters. The variety of surviving spectacle and eyeglass styles brings to the fore the role of the frame's appearance; overall aesthetic appeal was important to the prospective wearer. As shown in Chapters One and Four, spectacles and eyeglasses persisted in the range of a variety of high street stores including that of the jeweller and watchmaker who, although possessing technical skill, marketed the decorative features and characteristics of their stock. Moreover, in Chapter Four I demonstrated that the intended Victorian user was a savvy consumer driven by considerations of taste, trends and style as much as overall functionality and price. To purchase and wear a visual aid was to frame and fashion the face. But fashioning the face constituted more than the look of a frame. The choice and display of visual aid styles mediated meanings of seeing and blindness, and attitudes towards assistive technology that enhanced and restored Victorian vision.

This chapter starts by returning to Chapter One, and the beginning of our period, and finishes in 1904 to assess change and continuity in Victorian attitudes to partial sight and the wearing of spectacles and eyeglasses. Alun Withey has argued that a shift to wearing eyewear in public settings in the eighteenth century, as opposed to holding a frame to the face in a more private domestic space, altered the meaning and design of spectacle frames. While hesitant to denote a fashion in eyewear in the eighteenth century, Withey argues that spectacles 'shook off previous pejorative connotations'.[1] Indeed, the role of spectacles in fashioning the body and their ability to be publicly 'shown off' has its roots in the eighteenth century and the increasing importance of polite bodily display. Not desirable, but accepted, spectacles took on new, positive, social meanings, including learning and sagacity, as well as altering perceptions of what was materially possible for human vision and its correctability. But how complete was this shift? To what extent did new Victorian medical ideas affect the acceptance of spectacles as an assistive device or the device's potential to improve the lived experiences of its wearers? Did notions of disease and abnormalcy serve to rethread connections between spectacles and prejudice that eighteenth-century wearers and makers perhaps sought to sever? Conversely, did alterations in Victorian design offer an opportunity to more forcibly generate a 'fashion' in frame and lens styles or convey new social and cultural meanings? We cannot turn solely to medical understanding of vision and the eye to answer these questions. The increasing display of spectacles emerged alongside the broader role of clothing and accessories 'performing' on the metropolitan and provincial city streets.[2]

Claire L. Jones has identified a lack of scholarly attention given to the relationship between prostheses – their changing status and markets – and the conception of disability at a given time.[3] Withey, more specifically, has argued similarly, that we need to explore further how the use and perception of spectacles relate to normative attitudes to vision and its functionality.[4] As Jones and Withey contend, prostheses are not just 'artefacts'.[5] Spectacles, like other assistive devices, embody a range of different meanings and their purchase and use is influenced by the 'economic, social and cultural contexts in which they are designed, produced and promoted'.[6] Cara Kiernan Fallon's recent work, for example, connects the increasingly streamlined mode of production and design of the walking stick to an alteration during the early twentieth century in feelings associated with its use, from pride to shame.[7] As a result, the standardisation of stick design, as a more clinical, medical device, mediated the attitudes to, and experience of, walking impairments. Its materiality, the use of aluminium and rubber stoppers, altered the social and cultural meanings attached to the cane and reinforced newfound pejorative associations with its use. Withey, too, has demonstrated that the form of new assistive technologies has ideological sway, and can prompt recategorising of attitudes to, and experiences of, certain bodily conditions and/or differences.[8] Whether based on functional or non-functional considerations, the choices of material, size and use of body technologies – their very design – shapes the conceptualisation of disability, as well as the acceptance of, or stigma associated with, a device.[9]

A study of the walking stick and/or spectacles conflates scholarly work on body technologies and scholarly work on accessories. A large number of historical and interdisciplinary studies on accessories reveals that, in the Victorian world, similar items, such as the parasol or fan, operated in significant ways. Rather than 'marginal' or 'peripheral', often trivialised Victorian accessories performed subtle ideological work.[10] To return to Fallon's study of the walking stick, the Victorian stick's use was, on one hand, socially stratified and reflected users' economic choices; but, on the other, it was capable of influencing the experiences of ageing, disability and stigma.[11] Spectacles operated in a similar manner, but their significance was accentuated by their position on the face. As scholars have noted, in the eighteenth and nineteenth centuries the face was a powerful symbol of expression in an increasingly visual urban environment, where perambulating Victorian street spectators judged the appearance of passers-by.[12] Victorian anxieties about age, beauty and a lack of wholeness – markers on which spectacle use had a considerable influence – were brought to bear in the very public positioning of spectacles on the bridge of the nose. Conversely, Victorian visual aid users could choose a particular design to deliberately fashion their face in a manner similar to opting for a certain style of hat. Visual aids in this context were a tool to aid social mobility or defy gendered stereotyping that, on one hand, governed their popularity and, on the other, brought anxieties about women's role, masculinity or class blurring to the fore. It would be tempting to identify this solely as a problem or preserve for the middling and upper classes. To a certain extent, such a perspective is correct, but this chapter will also consider the increasing democratisation of the visual aid and the existence of second-hand markets with cheaper, mimicking styles that operated lower down the social scale.

As Jones hypothesised, alterations in the status and use of a prosthesis influence the conceptualisation of disability. In this chapter I explore this argument from two angles: first, the ways in which the very utility of optical lenses altered people's visual state from 'blindness' to 'seeing'; and second, the way attitudes to this type of assistive device shaped their acceptance. I have opted for the term 'fashioning' in this chapter to bring these two narratives together. First, fashioning – to form, mould, shape – reflects the ways in which optical lenses were chosen to suit people's eyes, to alter their visual state and the terminology attached to it, a process contingent on a variety of factors.[13] Second, to take fashion in its more literal sense – as 'an embodied phenomenological practice that draws distinctions of gender, class, age and status and that can express assertion and subversion, as much as docility or submission'[14] – in the second section I employ the idea of 'fashioning' the eye to reveal the way in which the display of certain styles of visual aid asserted and subverted dominant narratives of disability, gender, age and class. The very materiality of Victorian spectacle and eyeglass frames increased both usability and assimilation of the devices, which in turn influenced the experiences and perceptions of 'disability' as well as the parameters that separated stigma from acceptance. Medical intervention and terminology did influence the association of spectacles with disease, age and 'ugliness'. However, while an incomplete process, the growing utility of spectacles did much to challenge medical intervention and medical terminology by generating perceptions of visual aids as a useful device that, unlike its eighteenth-century counterparts, could be marketed as mainstream. Spectacles therefore offer a useful lens through which to view the ways that an assistive device can be culturally assimilated. Such assimilation was messy and contradictory.

Attributions such as attractive, ugly, valued, dismissed, normal, abnormal, stigmatised, desirable, depended on the age and gender of the wearer which, in turn, could imbue and convey a range of meanings among wealth, position, ageing, belonging, disease and defectiveness. As I will demonstrate, the adaptability of spectacle and eyeglass frame design meant that a range of styles emerged in response to the conflicting attitudes of a diverse usership, from the bold and outlandish to the modest and discreet. Such variation was not necessarily class-based but, unlike previous studies, this chapter will explore how lower visual acuity was often associated with the upper and middling classes who engaged in this market more forcibly, rather than the poor living conditions of the lower classes.[15]

In her study of artificial limbs Julie Anderson contrasted the increasingly standardised provision of artificial limbs with the individual experiences of artificial limb users.[16] It is important to not generalise the way Victorians perceived spectacles, or how their use influenced the experiences and conceptualisation of disability. While visual acuity was standardised, people's lives were not and the use of spectacles and the introduction of vision standards shaped their experiences positively and negatively, and in varying degrees. This chapter therefore has drawn upon a variety of sources to piece together the nuance of lived experience. To first look at how spectacles shaped the conceptualisation of blindness and people's experience of vision loss and partial sight, I use medical case accounts, individual reflections in personal and popular literature, opticians' cashbooks and letters. In the second section I then turn to popular literature – the vessel that enabled advertisement, fashions and attitudes to circulate – material evidence from spectacle and eyeglass frames and imagery to assess the way in which visual aids were represented. Spectacles 'restored' the way in which Victorians could view the world, accessorised an outfit and had the power to embody or project a person and their identity, yet, at the same time, their use in these contexts could be simultaneously accepted, emulated, satirised and stigmatised. Accordingly, Victorians fashioned their eyes with a variety of frames and styles to suit their own needs in a market often driven by taste and display.

'And thus literally gained an eye': spectacles and the conceptualisation of blindness

Case I – In April 1852, I was consulted by Anne Spencer, aged 23. She stated that she had always been blind of her left eye, and being obliged to work much with her needle had during the last few months found the sight of her right eye impaired from the labour thrown on it… Examination failed to detect any imperfection in the left eye, the pupil acted vigorously and the organ seemed in all respects sound and healthy… At three inches, she could see with the left. On trying her with glasses it was found that No. 4 Myopic brought the left eye into focus with the right, and, to her great surprise she discovered that with the supposed blind eye she could read with perfect comfort, assisted by that glass. She was directed to have a spectacle frame mounted with a suitable lens, the right circle being black, and thus literally gained an eye.[17]

William White Cooper, an ophthalmologist based at the North London Eye Institution and later St Mary's Hospital, introduced the case of Anne Spencer at the start of the second edition of his text *On Near Sight, Aged Sight, Impaired Vision and the Means of Assisting Sight* in 1853. Such a case brings to the fore a central question regarding the use of spectacles and the refashioning of the categories of blindness: did Victorian spectacles really work? While being careful not to present a whiggish narrative of progress, Victorian spectacles were certainly capable of improving people's vision and offered a newfound chance, in Cooper's words, to 'assist' visual capacity and sensory experience. Parallels can be drawn between the invention and adoption of new technologies and the classification of other bodily conditions. Coreen McGuire's work on hearing loss, for example, has demonstrated the ways in which telephony affected categories of hearing. Improvements in the use of technology altered the descriptors used to describe a person's level of hearing from 'extremely deaf' to 'hard of hearing', even where the overall level of hearing remained unchanged.[18] The adoption of spectacles by the early Victorians similarly altered the descriptors used in, and the classification of, blindness and partial sight. Cooper highlights that Anne had always thought she had been 'blind' in one eye when, in fact, she was only short-sighted. Such alterations were so profound that she was described as having regained her 'eye'. Using spectacles, Cooper was able to transform her 'blind' state into one that was able to 'see'. In the second case of his text, the use of lenses was not quite so transformative. Nevertheless, Miss H, aged 14, 'had never seen' with her left eye but, through use of concave lenses for her short-sightedness, found that with practice and gradual use 'rapid improvement' took place.[19]

In reflecting on the incidence of short-sight, Cooper argued it was often discovered by 'accident' by those affected. His description of a conversation with a patient also indicates how people's own experiences alter the way they view and measure both their own, and others', sensory capacity. On 'looking through a glass for the first time', the patient spoke of his 'pleasure' as though he had 'acquired a new sense' and that, prior to his own use of lenses, he had presumed others' vision to have been 'extraordinarily acute'.[20] Such personal insight serves to reinforce the care that must be taken to avoid determining sensory experience or the efficacy of a type of assistive technology as transhistorical. Cooper was writing at a time when Donders' work on the refractive condition of the eye and the diagnostic use of the ophthalmoscope – discussed in Chapter Two – was in its infancy. The restorative properties of lenses were an object of marvel for users and dispensers alike. Conversely, a host of more complex refractive conditions – often grouped under the term 'asthenopia' – had a bleak prognosis. One of the founders of ophthalmology in Britain, William Mackenzie, wrote that 'in many cases it is our duty to declare the disease incurable'.[21] As discussed in Chapter Two, a decade later Donders published how cases of, for example, the newly identified hypermetropia, which were treatable with lenses, had typically been misdiagnosed as asthenopia. Such alterations in how lenses could be used to 'restore' vision reclassified the permanency of certain forms of Victorian 'blindness', of the possibility of curing them, and thus what it was to be partially sighted. In contrast to other forms of technology or prostheses, the terminology used by patients or practitioners to describe the effects of spectacles reveals the transformative efficacy of lenses as an assistive technology. While categories of hearing could alter from 'extremely deaf' to 'hard of hearing' and an artificial limb might offer the opportunity to disguise a lack of 'wholeness', 'blind' patients 'recovered' their sight.

Jaipreet Virdi, in her analysis of the medicalisation of Victorian deafness, argued that 'the concept of cure itself is vexed' because, among other factors, 'the cure is only effective when the aid is in use'.[22] A 'cure' for Victorian partial sight or blindness was equally vexed. Like ear trumpets or other acoustic aids, spectacles only 'cured' when they were in use. Moreover, their overall effectiveness was contingent on a variety of conditions that have been tracked in the preceding four chapters. Fashioning a person's vision and the efficacy of optical lenses depended on improvements to the frames and/or lenses, eye examination, diagnostic technologies, knowledge of the eye and the quality of the device. A book by Francisque Sarcey, entitled *Mind your eyes!* and printed in London in 1886, gave a vivid account of his personal experience of being able to see fully for the first time:

> My father was master of a boarding-school... one day, for fun, I got hold of the large silver spectacles my father used to wear and put them on my nose as children will do in play. That was fifty years ago; the sensation I experienced is still vivid in my memory. I uttered a cry of astonishment and delight. Before that day I had never seen the vault of foliage which arched over my head, except as a large green compact sheet, through which no light penetrated. All at once I saw with surprise, stupefaction and rapture that there were openings in this dome through which light penetrated... what astonished me most and caused an enchantment which I cannot even now speak of without emotion was that through some holes in the foliage I suddenly perceived far away little bits of the blue sky. I clapped my hands and was in ecstasy. I was mad with admiration and joy. I could not rest until they gave me a pair of spectacles.[23]

This emotional response is not dissimilar to individual reflections on wearing lenses for the first time that have been recounted in my own lifetime, or my own personal recollection of being able to see the individual hairs on a person's head for the first time. Sarcey, too, points to his newfound visual acuity when looking at foliage that had previously been a 'large green compact sheet'. Writing from his perspective in 1886, Sarcey describes his former condition as an 'infirmity', and he subsequently encouraged his readers to get their vision appropriately tested. But how precise was Sarcey's vision? He had, after all, only put on his father's spectacles and his account pre-dates the standardisation of the vision test. Nevertheless, in popular literature, too, spectacles were lauded for their restorative properties. In 1876, contemporary literary critic Richard Hengist Horne wrote in his popular treatise on eyeglasses in *Fraser's Magazine* that spectacles are 'the real second sight', 'at once the rejuvenescence' and 'the preservative of the most important organ'.[24] Did spectacles warrant Victorians' 'curing' and 'preserving' claims?

In May 1871 a column called 'Table talk' in a weekly periodical suggested, in a brief account of their history and international use, that 'spectacles are worn by so many people nowadays that we are inclined to wonder how former generations managed to get on without them before they were invented'.[25] While not the intention of the column's account, this commentary emphasises important points that we need to consider in assessing the way spectacle use altered the conceptualisation of blindness. Correlating the simple invention of spectacles, mass-spread use and its subsequent inestimable benefits is too simplistic; adjusting lenses to a person's vision required a complex process of examination and adaptation. Second, in the columnist's speculation that 'perhaps their eyes were better than those of the present short-sighted race of mortals' they noted that in the past people had 'not so many newspapers to trouble them as we have'. The article thus highlights that the measurement and conceptualisation of sensory and physical impairment – and the subsequent fluctuating status and experience of blindness – was contingent on the social and cultural context of any given time. Mackenzie, for example, detailed a host of 'asthenopic' symptoms typically broken down into difficulty reading, headaches, pain and fatigue when engaged in close work.[26] Similarly, a surgeon at the Ophthalmic Hospital at Southwark, John Zachariah Laurence, gave an account of a tailor who, on account of astigmatism, suffered 'a sensation of burning in the eye-balls, congestion and lachrymation, all of which came on after reading &c. for a quarter of an hour or five minutes'.[27] While the treatment for astigmatism and other refractive conditions associated with asthenopic symptoms was relatively new, so too were the number of people engaging in activities that tended to trigger the symptoms.

The standardisation of the vision test did not automatically standardise a person's sensory experience by 'curing' all non-adhering visual capacities or automatically altering the diagnosis and treatment process to a recourse to lenses. The 'rejuvenating' effects of lenses were also contingent on the ability to correctly adapt and fashion them to a person's vision needs. While contemporaries celebrated the immeasurable benefits of spectacles, even simpler cases of refractive vision error could be misdiagnosed and treated wrongly. Laurence, for example, reflected on the case of Marcella D., 43 years old, who had suffered asthenopic symptoms since childhood and had received no previous treatment with lenses despite the fact that her symptoms had so intensified over the 'last eight or nine years... as to utterly incapacitate her for all occupations that demanded close work of any description'.[28] Although Laurence diagnosed Marcella with myopia and his treatment with 'corrective' lenses was effective, she had previously consulted sixteen or seventeen oculists. A decade later, Christopher Smith Fenner reflected that several patients with myopia were often 'subjected to harsh treatment' that could have been remedied with lenses by the 'ophthalmic surgeon'.[29] He suggested that the general practitioner often misdiagnosed patients' eye conditions that were 'daily seen' in his own practice. Such comments highlight the very personal implications of the unregulated sale of spectacles – or their dispensing by those who could not conduct a thorough vision test. Outside a hospital context, opticians' accounts and ledgers reveal the ways in which customers mediated the selection of lenses and the return of frames and lenses that left them dissatisfied. A single or several pairs of different types of visual aid could be bought and then returned, suggesting that the customer transaction operated around a trial period.[30]

As Laurence's patient Marcella D. hinted, without spectacles, refractive vision errors could have significant economic implications and affect a person's mental wellbeing. The inconsistency of spectacle use meant that not all Victorians were immediately 'cured' of their partial sight. Cooper also reflected on a patient's being 'extremely depressed in spirits' because of his inability to study or use his eyes, which was of 'serious consequence to him'.[31] Prominent British ophthalmologist Robert Brudenell Carter described a case from 1877 where, prior to appropriate vision testing, a partial-sighted patient was thought to have had a brain disease and, as a consequence, their 'whole life' had been 'blighted'.[32] Indeed, a series of letters and medical texts suggest that eye strain and the incorrect use of lenses was still affecting people's lived experience at the *fin de siècle*. Several investigations focused on the relationship between headaches, ocular discomforts and a range of refractive vision errors. Both Simeon Snell, ophthalmologist and former president of the British Medical Association, and Sydney Stephenson, ophthalmic surgeon to the Queen's Hospital for Children in London, illustrate the utility of lenses in alleviating a range of symptoms, especially the headache.[33] Stephenson identifies the persistence of these symptoms because, as shown in Chapters Two and Four, of the relatively late uptake of the matter by the medical profession in the 1880s and 1890s. A headache could perhaps be seen as a minor ailment. However, Snell argued that a headache could cause 'constant or recurring suffering' that 'frequently so disables its victims that life becomes a burden'. Stephenson, too, delineated the fine line between disability and functional capacity in the case of Miss Alice B., aged 35, who, with spectacles, could lead a normal life; but without spectacles her physical symptoms would return with 'former violence'.[34]

Snell and Stephenson, in their case accounts, demonstrate the ability of spectacles to reclassify a person's 'disabling' sensorial experience. Crucially, Stephenson points to lifestyle as being a significant factor in assessing and measuring the effects of eye strain and partial sight. He argues that 'the small error of refraction that would certainly pass without notice in an agricultural labourer might give rise to serious discomfort in a hard-worked literary man'.[35] Stephenson's cases demonstrate that serious discomfort could mean retiring to their bed, being 'frightfully weak' or considering a career change'.[36] In all the cases that Stephenson presented, the patient's vision was restored using a visual aid. A series of letters from the late 1890s and early 1900s was collated as part of a complaint that was sent to Moorfields Eye Hospital, to demonstrate the benefits of spectacles and encourage their use in treatment practice. It details accounts of patients attending a London optician, Mr. A. Fournet, who was able to treat conditions that previous consultations at Moorfields Eye Hospital had diagnosed incurable, with lenses. While, on one hand, these letters hint at the rivalries in opticians and ophthalmologists' practice that have been discussed in Chapter Four, on the other, they provide a rare hearing of the patient voice. A father writing a letter on 29 November 1898 included the prescription issued to his daughter Mary Gray by Moorfields in November 1889, when she was aged 6.[37] He was advised that his daughter should 'not be brought up to any occupation which necessitates the constant use of her eyes for near objects, such as writing, reading or sewing' and, such was her nature of progressive myopia, that she should be 'taught as much as possible orally'. Having attended Fournet in 1891, the father claimed that his daughter had 'seen well' for distance and close work ever since.[38]

The father's account suggests that the delayed use of spectacles not only affected his daughter's physical condition, but also her lifestyle and prospects. In 1899 J. Maddocks similarly described how he had previously required 'a stick to guide myself in walking' and that he 'could not read even large print'. Maddocks spoke of the 'relief' he had felt from the proper adaptation of spectacles to his vision, having been previously 'given up' as 'incurable'. Nevertheless, he stressed that his late adoption of an appropriate visual aid resulted in his being unable to complete his term of office in the public service and obliged to retire with a full pension.[39] As scholars have noted, and as discussed in Chapter Two, a growing proportion of the Anglo-American economy relied upon a person's functioning visual acuity.[40] A letter from J. Masters in 1899 stated that the timely adoption of spectacles meant that he had been promoted and was no longer in 'fear of not completing my term of service for a pension on account of my eyes'.[41] In contrast, in explaining his own experience of work with partial sight another correspondent, George Kirby, reflected that his eyes were causing him to become 'more disabled from work as time went on'.[42] The use of the word 'disabled' is poignant. The term in this context highlights the value of a visual aid to the user, its ability to extend a worker's working life, and the disadvantages faced by those who might not have had access to them. It was not always straightforward, however.[43] A letter written by Miss F. Perry in November 1900, who described herself as a 'useful maid', documents the difficulties of finding lenses to suit her sight.[44] Attached to Miss Perry's letters were several receipts, which documented eight visits to A. Fournet between June 1896 and April 1900. The purpose of seven out of the eight visits was to alter or rework her prescription and, in a one-year period between 1897 and 1898, Perry's lenses were altered five times: in February, June, July, October and the following February. She attributed her declining health to her vision, which was affecting her ability to 'keep in work'. After successive consultations, Perry described her interactions with Fournet as a success; she regained 'single vision at once' and was able to continue her employment.

The relief of these writers is palpable and W. Rudland in a letter dated 1899 gives an insight into his emotional response to his improved personal circumstances. He described his previous treatment under two practitioners at Moorfields who eventually told him 'not to come anymore as they could do nothing'. Spectacles, too, were deemed 'of no use'.[45] His consultation with Fournet had a markedly different outcome. Rudland wrote that his vision had been restored so as to read and write and 'drive a horse and van in the thickest traffic'. In reflecting on the challenges previously experienced moving through a fast-paced, busy urban environment, Rudland stated that he could now leave the house with 'pleasure… whereas, it was a misery for me to get out before'. It is worth bearing in mind that accounts such as these had a purpose, one that was intended to support the practice of a local optician. Nevertheless, the patients do describe their experiences of interaction with notable ophthalmologists at Moorfields, such as Robert Brudenell Carter, and are willing to verify or have others testify to their experiences. The father of Mary Gray, for example, did not see any personal gain in his testimony and instead adopted the rhetoric of the medical practitioner, of wishing to ensure 'incalculable public benefit'.[46] It is in the surviving voice of the atypical patient, those with a complex, more extreme, case of refractive error, that the efficacy of spectacles for reconceptualising experiences of blindness is most profound albeit, as in the case of F. Perry, a lengthy, delicate adjustment. In these cases spectacles were a restorative aid to the wearer, one able to lengthen their working life and improve their mental and physical wellbeing. Spectacles did alter perceptions of functional visual capacity. But the idea of 'cure' was not always fully realised. To 'fashion' – or to form, mould, shape – new categories of partial sight and vision depended on access to, and the appropriate utilisation of, technology. Both the wearer and the medical practitioner review the efficacy of spectacles in their ability to overcome 'disability'. This process, as the letters of complaint reveal, could depend on the optician; the celebrated 'restorative' properties of spectacle and eyeglass lenses were contingent on a complex interplay of expertise that was not just medical in tone. In such instances of 'restoration', the prospective spectacle wearer was neither blind nor 'incurable'; they could see 'well' and participate in both recreational and occupational tasks expected in the Victorian world.

The desired, satirised and stigmatised frame: viewing visual aids from a Victorian perspective

In returning to Dr Dyce Davidson's observations from 1885 that I introduced at the beginning of this book, Victorians were acutely aware that the demographic of spectacle use had altered during their lifetime. Davidson identified that, traditionally, spectacles had been the 'first warning of declining age' and for younger wearers were 'at once the object of remark' akin to being 'afflicted with a deformity of body or limb'.[47] Within his 'own memory', Davidson identified a nine-fold increase in spectacle wear, particularly in young adults. Davidson was not the first to identify this growth in visual aids. Correspondents to *The York Herald and General Advertiser* in the 1840s sought to educate readers on 'unprincipled spectacle venders' because 'the use of spectacles by persons of various age has become so general… and the assistance rendered by them to many thousands of our fellow creatures so beneficial'.[48] By the 1880s the 'many thousands' had been rearticulated and reached the proportions of 'an epidemic'.[49] Just as Davidson discussed increased spectacle use in the context of educational overpressure, several articles emerged discussing the many benefits of modern society set against 'indications of a decidedly lower sight average'.[50] Lighting in streets and houses, large shops in public thoroughfares, new industries and education were all thought to have a bearing on spectacle usage and its growth. From this perspective, the personal responsibility to take care of one's own ocular health was stressed but spectacle users were presented as 'victims' of forces outside their control. Under the heading 'Victims of industry' in the *Westminster Review* in 1894, a writer called for occupational health intervention to ensure the working conditions of a variety of employments – printing, tailoring, shoemaking, banking and counting-houses – were fit to avoid 'partial or total blindness'.[51]

Claims of 'many thousands' of victims, or 'epidemics', in popular literature cannot perhaps be taken at face value. The development of mass production and the expansion in retailing sites for spectacles evaluated in Chapters One to Four does, however, suggest that there was a larger proportion of visual aid users driving demand between 1850 and 1904 than had been the case earlier. Letters and case accounts similarly evidence that a range of users existed, and that the demographic of spectacle users, as Davidson observed, was not confined to the more elderly. The cases of complaint to Moorfields Eye Hospital discussed in the previous section, for example, included several children receiving treatment for eye conditions with spectacle lenses or therapeutic methods.[52] Indeed, medical and opticians' case accounts include a broad cross-section of spectacle users, men and women, boys and girls, receiving treatment for a variety of refractive vision errors.[53] How were these wearers perceived? Were visual aids worn willingly? It is not possible to generalise attitudes towards spectacle usage across this broad demographic.[54] But it is possible to examine the variance and nuance of individual experience: the elderly wearer, the young woman conscious of her appearance or fashioning her dress, the young man trying to obtain work or adding an eyeglass to his appearance, the educated man or woman using spectacles to bolster their reputation and those who wanted to mask conditions otherwise visible upon the face. Whether accepted or rejected as a visual aid user in broader society, these groups of people adopted eyewear as a tool to improve their vision and/or convey a message to passers-by on the street, in the workplace and at events. Spectacles sit at the heart of a much broader interplay between technology, culture and the Victorian body.

Social and cultural responses to visual aid wearers and/or the use of spectacles and eyeglasses as an assistive device reflected the age, gender, class and context of the wearer. Nowhere is this more evident than in considering the cultural associations Victorians attached to spectacle use and age. The relationship between spectacle usage and old age was a historical and intimate one. For Davidson, spectacles had long been the 'first warning' of the onset of senescence. A column in a Saturday periodical twenty years earlier had likewise argued that the first notice of a decline in visual acuity is 'an epoch' wherein 'our independence, our freedom, our youth is gone'; we resign ourselves to 'henceforth... hold our paper at arm's length like the old fogies in *Punch*'.[55] Since the eighteenth century, the universality of spectacle use in old age was recognised but not fully understood.[56] Advancements in medical understanding as a result of the incorporation of the ophthalmoscope into medical practice identified the universal stiffening of the lens's accommodative – or focus-adjusting – power which made it harder to view objects or text close up. Nevertheless, the association between prosthetic use and age and bodily decline was of much broader significance for the Victorian.[57] Victorians were anxious about ageing and particularly the early, perceivable signs of senescence that appeared in the newly identified stage 'midlife'.[58] Scholars have noted that the proportion of older people in the Victorian population had in fact shrunk, but that cultural attention to the experience of ageing was far greater.[59] A growth in products to combat signs of ageing – advertised as either patent medicines or beauty products – emerged alongside the growing belief that bodily decline could be prevented.[60] As Ryan Sweet has demonstrated, the nineteenth-century prosthesis market drew upon this rhetoric and prostheses – here, artificial limbs, wigs and false teeth – were another method to disguise physical decline.[61] When prosthesis use was exposed, however, evidence of it had the power to age the wearer by association.[62] The visibility of spectacles on the face, therefore, often appeared to increase or draw attention to the apparent age of their wearer, regardless of their physical age. London opticians Thomas Harris & Son argued in 1839 that people 'dislike even the bare thought of using glasses, because, as it is said, they make them look *so old!*'[63] In the 1890s, despite a perceivable growth in the demographic of spectacle users, a popular account of

William Gladstone's visit to Midlothian in Scotland observed that his use of spectacles made him seem 'at least ten years older'.[64] The association of visual aid use with age, or as an accessory predominantly for the elderly, is also apparent in photographs between the 1870s and 1890s that depict children dressed up as their elder relatives. A photograph of a 'Little Grandma' from 1889, for example, depicts a girl wearing spectacles and a bonnet while seated and knitting.[65] Other photographs included similar depictions with accompanying captions: 'Now I'm Grandmamma', 'I'm Auntie Now' and 'Miss Giles Dressed as an Old Lady'.[66]

As the age markers used in the children's photographs suggest, the meanings attached to the onset of old age are biologically and culturally determined. Scholars have identified that the boundaries used to identify the categories 'midlife' and 'old age' are open to interpretation and contingent on a variety of factors.[67] Increased medical attention to the physical aspects of bodily decline identified the elderly as a cohort for clinical study and classified symptoms of senescence as either 'morbid' or 'normal'.[68] In the early twentieth century the Austrian-American physician Ignatz Leo Nascher, who was the first to propose geriatrics as a medical speciality, identified a distinction between natural ageing and pathological alterations.[69] Since 1850, ophthalmologists had utilised similar terminology; the process of weakening accommodative power in old age was deemed 'natural' and 'universal' and not categorised as a disease in the way that other refractive conditions, such as more severe forms of myopia, were. Culturally too, Victorians, while anxious and desiring to combat outward physical decline, did not have a monolithic view of ageing as an inherently negative experience.[70] Optimistic and pessimistic paradigms of ageing were influenced by the context in which a person was situated. Old age could represent dignity and wisdom and Victorians had powerful ageing figures embodying these values.[71] As an assistive device not intended to disguise, therefore, spectacles could act as a tool to communicate positive social markers associated with age. Doctors and clergymen, for example, were described in the *Hull Packet and East Riding Times* to be 'especially partial to spectacles' because their use by these groups was legitimised and able 'to lend gravity to their looks and enable them to pass for sages amongst the ignorant'.[72] As Melissa Dickson has shown in her study of doctors' professional dress, objects could be markers of professionalism and convey reliability.[73] Indeed, in the *Hull Packet* the writer estimated that 'in ninety cases out of a hundred' wearers 'don't require them' and that they are only chosen because of their associated markers of dignity, wisdom and intelligence.[74]

In contrast, culturally contingent factors of ageing intensified the stigma already attaching to visual aid use in certain cohorts and demographics. A young woman wearing spectacles was perceived differently to a young man wearing them or an elderly woman doing so. Women were considered to age faster than men and symptoms of physical decline were more severely scrutinised.[75] The literary critic Richard Hengist Horne, in his 'Eyes and eyeglasses' treatise mentioned earlier in this chapter, argued that young women tended to avoid visual aids because they were not 'graceful and becoming, and they certainly add something to the apparent age of the wearer'.[76] Horne identifies not only age but a want of elegance as a reason, in social situations, to avoid wearing a visual aid altogether, even when required. Men were not immune to criticisms of vanity. A statement in *The School Board Chronicle* in August 1889 argued that medical practitioners and opticians could be 'particularly severe on the men and women who suffer the inconvenience and danger of semi-blindness and sacrifice their eye-sight to the vanity which regards spectacles as unsightly'.[77] Nevertheless, contemporaries tended to excuse women from wearing visual aids because of its potential bearing on their marriage prospects. Horne, for example, argued that 'fair allowances' can be made for women, but 'no excuse for men'.[78] In his seminal work Donders stated that in some cases concave lenses need not be used even when they are required, and that 'women particularly have a right to be allowed some liberty in the matter'.[79]

Why might a leading ophthalmologist, whose work had global influence, grant women an exception? Concave lenses were for long distance and would have had a greater public function than the more private use of a convex lens for close work. Vanity and a desire to avoid public display could be explained by lack of aesthetic appeal or spectacle and eyeglasses' association with premature senescence. It also encompassed the desire to hide blemishes and portray an image of physical wholeness.[80] More severe refractive conditions of the eye that would have been treated with concave lenses and have required more permanent spectacle wear, namely myopia, were associated with disease and it was not possible to visibly distinguish between the 'natural', 'universal' need for spectacles in old age, and the newly diagnosed 'pathological' states of the eye that necessitated 'corrective' lenses. Simply observing a spectacle wearer did not make it possible to determine whether a young person was at the onset of midlife or whether the condition of their eyes was a sign of a more systemic degeneration. The hereditary nature of myopia was a cause for concern. Dr Bell Taylor, speaking on myopia and short-sight and their possible relation with education, drew upon the remarks of an 'eminent professor of ophthalmology' at Harvard University, Dr Williams. Dr Williams argued that 'very high degrees of myopia should be recognised as an infirmity deserving careful consideration before assuming the obligations of marriage'.[81] While Williams' views were not representative of the medical profession, practitioners found the hereditary nature of myopia alarming. As shown in Chapter Two, myopia could have a bearing on a person's overall physical health, particularly during childhood, and fed into broader concerns about degeneracy and the health of the nation.

'Completeness' was not a solely female marriage requirement, and recent studies have shown the importance of physical prowess for men in navigating anxieties about age, disability and physical decline.[82] As Ryan Sweet identifies, this could intensify in relation to work, where physical difference could affect a man's ability to obtain employment.[83] A lecture on 'Our eyes and our industries' reported in the *Freeman's Journal* in 1887 detailed the case findings of Priestley Smith, an ophthalmic surgeon at the Birmingham Eye Infirmary. Smith exposed the double-bind affecting male spectacle wearers because of the 'ignorant prejudice' that 'able and skilled men' could not 'do their work without glasses, and could not get employment if they wore them'.[84] The implications of such a prejudice could be severe and men 'thrown out of work' were also thrown 'into pauperism'. The objective of Smith and Dr Arthur Benson, who delivered the lecture – to impress that wearing spectacles did not mean that the eyes 'were breaking down' – indicates that stigma towards spectacle usage and its association with degeneracy could circulate outside medical control and fed into broader concerns about the deterioration of vision being inflamed in popular literature. It was also bolstered by a range of factors, aside from medical opinion, including the influence of capitalism and the growth of industrialisation which, as discussed and critiqued in Chapter Two, laid emphasis on bodily (in)capacity.[85] Yet Smith's focus on the social and economic disadvantages that men could face was more unusual because, in the context of spectacle wear, it was women who were foregrounded in male medical writing. Women's social and life prospects were reduced, and women could be seen as more susceptible. Associations, for example, were drawn between women's susceptibility to other conditions such as 'imaginary neuralgia' or 'chronic headache' that were also symptomatic of refractive vision errors.[86]

As a central feature of the face, spectacle wearers had two potential options to combat the varying response to eyewear use: the concealed, discreet frame that sought to disguise its use and the bold, decorative frame that sought to position spectacles or eyeglasses as an accessory. The form and materiality of spectacles – like other assistive devices – influenced the experiences of impairment and social access.[87] As historians of prostheses have noted, one of the primary functions of prostheses was – and remains – to enable its user to 'pass' and render any lack of conformity invisible. 'Passing' is a term used to describe an attempt to conceal social markers of impairment or avoid the stigma of disability. Patent applications from the 1880s sought to disguise spectacles in a variety of everyday items for public use, including the handle of an umbrella, parasol, walking stick, whip or fan.[88] However, Victorians also could seek discretion in frames to be worn on the face. The rimless frame was first made in spectacle form between c.1825 and 1840, and the earliest rimless frames in the London Science Museum's collections date from 1844.[89] The term 'rimless' was not adopted by Victorians and instead spectacle retailers opted to market these frames as 'invisible'. 'Invisible' designs from the 1870s proliferate in both the London Science Museum's collections and in contemporary advertisements in popular literature.[90] Parallels can be drawn between this style and the desire to produce more realistic prosthetics in this period.[91] 'Invisible' styles were driven by responses to physical difference and a desire to hide defects as much as they were driven by improvements in manufacture that made different styles and devices possible. In a practical sense, invisible styles were flimsy and liable to break. Nevertheless, they could disguise or even hide eyewear use and therefore reduce or avoid the attention spectacles drew to the face, whether because of their association with unattractiveness or as an indicator of visual abnormality or age.

Just as rimless frames were emerging on the spectacle market, in 1848 the miscellaneous column of *The London Journal and Weekly Record of Literature, Science and General Information* informed its readers on the 'Physiology of vision'. It argued:

> The desire to conceal from the world any imperfection which wounds our self-love, is inherent in the human heart, and leads to all sorts of artifices on the part of those who, by natural conformation, advancing years, or other causes, suffer from imperfection in their vision. Thus it is, that persons prefer to use an eyeglass, others reading-glasses, in lieu of spectacles.[92]

The column preceded to condemn wearers of a single eyeglass or reading glasses because of their instability and the 'trying' nature on the eyes of such devices. As Chapter Three has demonstrated, eyeglass frames did improve across the Victorian period and, for certain activities, were eventually accepted by opticians and medical practitioners alike. Yet rather than necessarily evidencing social acceptance, contemporaries revealed that eyeglasses and elaborate designs helped 'conceal' 'imperfection'. Later in 1848, a description in *The Athenaeum* of a person using a pair of double eyeglasses argued that people carried them 'so gaily, you would have hardly known it was spectacles in disguise'.[93] In this way, spectacles are another piece in the toolkit of those turning to commodities to mask or disguise their body or social standing in what a series of articles in *Fraser's Magazine* in the early 1850s coined as the 'Age of veneer'.[94] The styling of eyeglasses offers a different perspective of users' attempts to mask their assistive device and is, for example, comparable to Stephen Mihm's identification of 'lifelike' limbs designed for the genteel operating in 'polite society'.[95] Instead of masking their partial sight, the middling classes could adopt eyeglasses to navigate their marriage, employment and social prospects. The use of eyeglasses in this way is in opposition to the desirability of invisibility in Victorian prosthesis design. Eyeglasses were a 'show' item that sought to draw attention to the frame and could fully supplant their functional purpose; a multifaceted accessory that, like the fan or parasol, was capable of shaping attitudes towards the functional body. Indeed, contemporary commentary suggests different styles of frames for public and private use persisted. In 1891 a miscellaneous column headed 'Over the teacups' commented that old age is made more 'cheerful' because the optician can 'prescribe them with eyeglasses for use before the public and spectacles for their hours of privacy'.[96] In this vein, eyeglasses perhaps fit more neatly within the historical context of the Victorian beauty and cosmetic market where scholars have traced a deliberate fashioning of the body.[97] The appearance of the frame was equally important to the Victorian buying a cheap frame and those who could afford to choose a frame based on taste and aesthetic preference. Retailers, for example, used terms such as 'elegant' or 'handsome' across their price range in advertisements in the popular press in the 1830s, 1840s, 1850s and 1880s.[98] Advertisements and material evidence reveal that there is little to suggest that spectacle design was overtly gendered in its bid to become attractive. It was more unusual for a retailer to target particular designs on 'ladies' or 'gentlemen'. When they did, opticians stressed lightness as a key feature for ladies, or other potentially desirable traits such as not disarranging the hair.[99]

Whether we can ascribe the popularity of particular materials or styles of visual aids to 'fashion' is difficult to say. Certain optical firms were reputable and well known – for example Dollond and Dixey – but there are marked differences between this fact and, for example, the rise of commercial eyewear brands that has been explored in the twentieth century.[100] Moreover, the fashion for eyeglasses was not universal. Surviving rimless eyeglass frames suggest that not all wearers wanted a bold, public style and sought a discreet frame.

Nevertheless, while Withey argued that there was not a 'vogue' for spectacles in the eighteenth century, Victorians did observe a change in the way in which visual aids were worn. Materially, this is evident on the frames of some of the visual aids in the Science

Fashioning the eye and seeing, 1830–1904   229

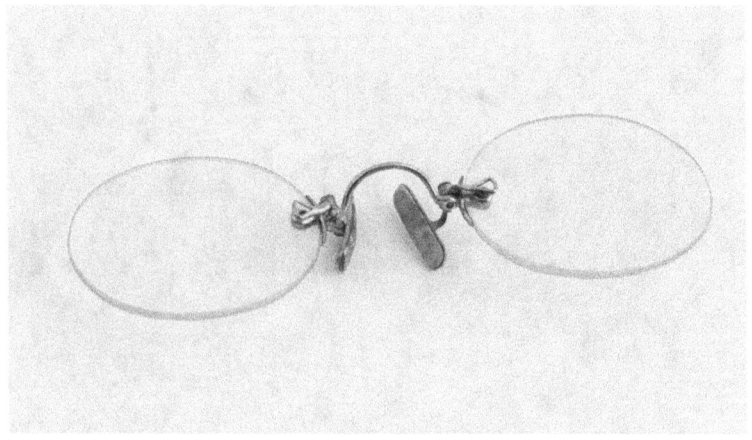

Figure 5.1 A pair of rimless 'finger piece' gold clip-eyeglasses, dated to c.1890 and indicative of certain eyeglass styles intending to be discreet.
Source: Science Museum Optics collection, 1921-323/379. With permission from SSPL/Science Museum.

Museum's collections. The elaborate etchings and use of pearl and luxurious materials, for example in Figure 5.2, show that, even if worn for functional reasons, the visual aid could act as an accessory to signify a person's status, position, or style.

An article on 'Spectacles' from 1877 regarded spectacles and eyeglasses 'much in the same light as diamond rings and patent-leather boots'.[101] With this in mind, eyeglasses were often described in popular literature as being used by the 'aristocracy' to observe 'the working men', 'or hold themselves as creatures apart'.[102] Here the eyeglass, like other accessories of the time, was a tool for class performance.[103] Depictions of spectacle usage circulating in the upper and middling classes offers a contrasting perspective to previous studies of nineteenth-century vision in which scholars argued that lower visual acuity was associated with the lower classes and poorer living conditions.[104] Medical practitioners in the early Victorian period and the popular press in the later Victorian period do assess the relationship between nutrition and the environment of the lower classes and its bearing on

Figure 5.2 A slightly earlier style of gold folding eyeglasses, with an ornate handle and detail on the eyepieces.

Source: Science Museum Optics collection, 1921-323/286. With permission from SSPL/Science Museum.

vision defects, particularly in relation to schooling and domestic conditions.[105] However, in reality, medical practitioners observed that lower visual acuity was not symptomatic of poorer living conditions. Indeed, medical practitioners found that the upper classes were more susceptible to 'abnormal' refractive conditions of the eye because of their tendency to occupy themselves with activities that involved indoor close work.[106] Late Victorian ophthalmologists were heavily influenced by F.C. Donders' discussion of the relationship between class and the use of visuals aids. Donders argued that 'position in society has a great influence' and the living conditions of the upper classes had a greater bearing than the environment of those in poverty.[107]

'Position in society' had a comparable influence on the style and portrayal of spectacle users. By the 1880s, both men and women in the middling and upper classes partook in the 'fashion' for eyeglasses on both sides of the Atlantic.[108] Nevertheless, while as markers of wealth or appendages to a person's outfit and style, fashions in eyeglasses reveal the uneven ways in which visual aids were assimilated in British Victorian society as well as how they could upset the status quo. Certain fashions and deliberate stylings were celebrated while others were denigrated. In 1884 *The Preston Guardian* observed the latest 'fashion' for ladies, who adopted eyeglasses to impart 'an extra look of interest to them in the eyes of the opposite sex, who don the sight preservers with a similar idea'.[109] The article gave women agency and presented eyewear use as a decision based on both functional and aesthetic considerations. Indeed, recent scholarly work on women's engagement with fashion and consumer culture highlights that the purchase or use of an object is a 'wilful act'.[110] An article on spectacles in 1880 similarly suggested that ladies chose to wear spectacles 'exclusively in the company of their own sex' and an eyeglass or *pince-nez* 'in general society'.[111] However, the article argued that glasses, of whatever style, are 'apt to give a semi-masculine, semi-scholastic, semi-clerical appearance to female wearers, which is not particularly prepossessing'. Such an account reveals how positive social markers such as sagacity and learning were not universally borne by the eyeglass or spectacle wearer. Yet it also reveals ways women could conscientiously adopt a prop in a climate where the use of commodities to improve one's look, or convey a meaning in society, was expounded in Victorian etiquette manuals. Susan Hiner in her study of Victorian accessories draws upon the legal definition of an 'accessory' to explore the ways they were used by women to aid and abet their social standing.[112] Here, spectacles become an 'accessory' in the legal sense of the word by accompanying and aiding women in their subversion of accepted social and gender stratifications and in their wielding of power. In contrast to a wearer's decision to disguise, or the importance of modesty in the design of women's prostheses, between the 1870s and the 1890s jokes foregrounded women's use of visual aids to comment on their association with masculinity and a potentially threatening intelligence.[113] A British satirical description of an American 'Boston girl' who wore 'a double eyeglass', for example, specified that 'her lightest small talk is of palaeontology; and her highest idea of relaxation is to attend a course of lectures on the polarisation of light'.[114] Another joke in the column 'Odds and ends' in *The Illustrated Police News* also drew upon eyewear and the Boston girl:

You don't seem as friendly to that Boston girl as you were. 'No; we were out riding one evening, and I put my arm around her and told her I loved her.' 'Did she refuse you?' 'No, not exactly; she looked at me over her eyeglasses and remarked, "Mr Jones, do you know what a protoplasm is?"' 'What did you say?' 'Say!' replied Mr Jones, with deep disgust, 'What could I say? I never was in Boston.'[115]

Similar jokes to satirise and destabilise a woman's position when discussing her 'rights', social freedom or education were peppered throughout the Victorian press in the 1890s.[116] These jokes suggest that, on one hand, the campaign for women's rights could be undermined by using visual aids, which could be used to represent women as masculine and make bookishness appear eccentric. On the other hand, scholars have shown how illustrations of women in overtly masculinised roles can be seen as representations of male social anxiety at the advancement of women's position.[117] Moreover, Lynda Nead has highlighted that satirising the size of Victorian women's crinolines reflected male anxiety at women taking up too much physical space in the public world outside their home. Indeed, she offers the argument that women were active agents in their dress as a counter-narrative to their being slaves in the trappings of a fast-paced consumer culture.[118] Eyeglasses jokes support this conclusion and indicate how women may have adopted and controlled their own use of eyewear to challenge social stratifications. Indeed, spectacles and eyeglasses could be used to make a woman appear impudent. An article reflecting on 'feminine warfare' described the use of an eyeglass by a woman, who was not short-sighted, but used her eyeglass as a means to portray 'sublime contempt' to her fellows.[119] Rather than being presented as passive, women's use of eyewear in these instances supports the view that women were as much spectator as spectated in Victorian society.

Before the more widespread fashion for double eyeglasses in the 1880s, the decorative visual aid had long appeared as an accessory to a woman's outfit. The single eyeglass, for example, featured as part of the latest ladies' fashions in *La Belle Assemblée; or Court and Fashionable Magazine* between 1818 and 1825, a magazine that has been considered an important example of its time in the way it appealed to women's tastes for both fashion and current affairs.[120] In 1823, for example, a description of a 'walking dress' from 'Fashions for November' stated that 'a gold chain, with a large perspective eyeglass, is generally adopted with this dress'.[121] When accessorised or worn for display by Victorian men, an eyeglass or eyeglasses became the subject of satire and debate. For men, the eyeglass had become synonymous with the dandy and satirists could use eyeglasses to promote an image of effeminacy. In 1871 a satirical image in *Fun*, entitled 'A Man of Standing', alludes to the contrasting association of an eyeglass with status or 'standing' by its user, and the perceived delicacy of the male eyeglass wearer:

> *Eyeglass:* Rather a stiff breeze this to stand out in, Boatman.
> *Sealegs:* Stiff, is it? If yer can't stand out in a breeze as stiff as this, what can you stand?[122]

Here, the strength of a man entitled 'sealegs' is at odds with the eyeglass wearer, who is struggling to remain upright in what is presented as a light breeze. Just as the eyeglass could be perceived negatively for making a woman appear masculine, groups of men who chose to wear an eyeglass could be considered effeminate. Whereas women's apparent need for them for functional reasons was criticised – and their use as an accessory was not – men's need for them was not criticised and their use as an accessory was. The choice of title, 'A Man of Standing', also reflects deeper anxieties about the blurring of sartorial and class boundaries in the second half of the nineteenth century. While we have seen eyeglasses being used by the aristocratic to view their inferiors, the dress of the dandy was increasingly appropriated by middle-class men. Such satire thus reflects a broader anxiety about emulation through the utilisation of mass production of clothes and accessories, which created what Shannon Brent has termed 'class confusion and conflict'.[123]

The potential use of the eyeglass for non-functional purposes, and the adoption of cheaper styles, further fed into cultural anxiety about the use of commodities to perform and the subsequent hollowness, and vulnerability to trickery, of such practices.[124] This could create mistrust and further stigmatise men who did wear a visual aid for functional reasons. An account in the *Dublin University Magazine* from 1877, by a man who dubbed himself 'the London hermit', reported that people who were short-sighted could be judged as imposters or pretenders. He described a variety of situations in which he had been dismissed or ridiculed by individuals who had presumed that he was wearing spectacles for show.[125] Such accounts highlight that the use of eyeglasses for show, prior to their increasing fashionability, had traction. Indeed, while it is impossible to know how many people wore a visual aid unnecessarily, British popular literature had long commented on the trend's existence. In 1850, for example, it was argued that 'plain glass is the most harmless contrivance for those who insist upon looking through a window to avoid the simplicity of ungarnished eyes'.[126] On the eve of the emergence of the term 'fashion' in association with eyeglass use, the number of pretenders had supposedly reached sizeable proportions. In 1873 an article headed 'Social statistics' argued that:

> [i]t has been ascertained, by personal confession, that out of a thousand very elegant young gentleman afflicted with the monomania of wearing an eyeglass, only five were in reality afflicted with an actual infirmity of sight.[127]

While it is likely that such accounts are only the perceived extent, rather than the reality, of this form of eyeglass wear, the use of eyeglasses or spectacles by 'pretenders' indicates that they were not just perceived as a functional, assistive device. Visual aid usage was part of a much broader concern about people's relationships with objects.[128] A 'handsome' or 'fashionable' frame did not improve the frame's functionality, but it did improve its ability to be displayed and convey the status, intelligence, style or social positioning of its wearer. In the 1890s, utility and appearance took on the same level of importance in both medical and optical texts. Purchasing power did have a bearing on the meanings that a spectacle or eyeglass wearer could portray. However, as this book has demonstrated, those lower down the social scale had access to increasingly mass-produced products, second-hand markets and the display of elegant, but cheap, styles of frame. Moreover, while functionality may not have been prioritised by the lower classes, Sweet has identified that such prioritisation did not mean that the appearance of the device was not important to prosthesis users.[129] A wealthy member of society flaunting their luxurious eyeglasses on the street – whether required or not – was a fundamentally different mode of dress to the worker donning a pair of more modest steel spectacles to meet the visual requirements of their employment. But, for spectacle users of little means, an elegant frame was not unattainable or, indeed, thought unnecessary. In 1898 an article on frames in *The Optician* impresses that the quality of the frame does not determine its overall appearance on the face. It argued that 'the finest finished and tempered frame' was 'worthless' when it did not fit or 'look well'.[130] Not everyone could afford the 'finest finished' frame but they could achieve a good fit with malleable steel-wire frames. Indeed, the cheap, steel-wire utilitarian frame discussed in Chapter Four was marketed as 'elegant' and stood in marked contrast to the older, more cumbersome, styles of the early nineteenth century.

Conclusion

To 'look well' wearing a visual aid was a subjective opinion. A person's use of eyeglasses and/or spectacles could be accepted or rejected and stigmatised dependent on the age, gender, wealth and context of its wearer. On one hand, wearing them could evoke feelings of shame and portray an image of weakness or premature old age. On the other hand, it could bolster a person's reputation by signifying wealth and intelligence, as well as offer a means to disguise an area of the face. Often, therefore, care was taken by the user and the dispenser in suiting the type of frame to a person's face based on the frame's intended purpose. In 1898, William Smith Baxter, a qualified optician and member of the British Optical Association, stressed that his 'comfortable' spectacles and eyeglasses were chosen to suit the individuality of a person's appearance and 'adorn the face', even for those with the most 'fastidious tastes'.[131] Similarly, Sheffield-based opticians Chadburn and Son argued in the 1890s that the 'material, and to some extent the form of the frames are a matter of taste'. The variety of styles, retailing outlets and manufacturing possibilities for visual aids created a device that could be adapted to the person and situation as required. Indeed, the relative assimilation of spectacles can be measured by the way they were used to hide or mask other conditions. The utilitarian – not decorative – visual aid could facilitate, for example, the attachment of artificial noses or conceal early forms of assistive hearing devices.[132] More broadly, an article in 1893 in *The Optician* headed 'Spectacles for cosmetic effect' reflected that spectacles could improve the appearance of patients affected by a range of eye conditions. It argued 'neatly fitting' spectacles could render the 'unsightly' appearance of the eyes of the blind 'less lugubrious' by either magnifying the eye to the 'right' size or masking them with coloured lenses. It also highlighted how spectacles could aid in the treatment of epicanthus to 'smooth out the offending fold of skin'.[133] In this instance, the 'success' of the eyeglasses or spectacles as an assistive device was measured by its ability to render the other condition invisible through foregrounding the *visible* visual aid.[134]

This chapter has identified the ways in which the design of spectacles, and their changing status and market as an assistive device, refashioned the Victorian meaning of blindness and partial sight. On one hand, certain conditions that could be 'restored' with lenses transformed people's state from one of blindness to being able to see. On the other, the varying responses to spectacle use created an interwoven, contradictory and complex set of attitudes towards partial sight that were formulated outside medical control. The Victorians accelerated a lot of the trends that Withey identified in the eighteenth century. Fashioning the face and the way a visual aid could be used as a tool by its user – regardless of whether it was accepted or rejected more broadly by others – brought eyewear into a new 'vogue'. It was not a straightforward assimilation. Users sought a variety of styles in a bid to 'pass' or 'belong' in a social situation. 'Passing' is contentious in disability studies and among disability activists. Nevertheless, Jeffry A. Brune and Daniel J. Wilson have rightly noted that it is more fruitful to explore and challenge 'the ableism that compels people to pass rather than blame the individuals who choose to do so'.[135] Partial sight and wearing spectacles could be stigmatised. This stigma took a variety of forms from affecting a person's marriage or employment prospects to presenting the visual aid wearer as the subject for a range of jokes depending on their age, gender or social situation. The uniqueness of the spectacle and eyeglass frame lay in the numbers of prospective users and in the ability to produce an assortment of different styles of frames to suit the users' needs: the elegant, but affordable, steel frame; the discreet 'invisible' frame; and the bold decorative frame. Spectacles and eyeglasses were not just functional 'things', their very materiality embodied a variety of meanings – from positive markers of wealth, style, intellect and dignity to more negative markers of age and perceived physiological weakness – that its users sought to simultaneously mask and display. To fashion the eye was to offer a visual marker of individual expression that could reinforce and subvert dominant narratives of disability, gender, age and class in Victorian Britain.

Notes

1 Alun Withey, *Technology, Self-Fashioning and Politeness in Eighteenth-Century Britain: Refined Bodies* (Basingstoke: Palgrave Macmillan, 2016).
2 Christopher Breward, 'Masculine pleasures: metropolitan identities and the commercial sites of dandyism, 1790–1840', *The London Journal*, 28:1 (2003), 60–72; Shannon Brent, *The Cut of His Coat: Men, Dress and Consumer Culture in Britain 1860–1914* (Athens, OH: Ohio University Press, 2006), Catherine Waters, '"Fashion in undress": clothing and commodity culture in *Household Words*', *Journal of Victorian Culture*, 12:1 (2007), 26–41; Janine Hatter and Nickianne Moody (eds), *Fashion and Material Culture in Victorian Fiction and Periodicals* (Brighton: Edward Everett Root, 2019).
3 Claire L. Jones (ed.), *Rethinking Modern Prostheses in Anglo-American Commodity Cultures 1820–1939* (Manchester: Manchester University Press, 2017), p. 2.
4 Withey, *Self-Fashioning and Politeness*, p. 93.
5 Jones, *Rethinking Modern Prostheses*, p. 5; Withey, *Self-Fashioning and Politeness*, various; see also, Graeme Gooday and Karen Sayer, 'Purchase, use and adaptation: interpreting "patented" aids to the deaf in Victorian Britain', in Jones (ed.), *Rethinking Modern Prostheses*, p. 39.
6 Jones, *Rethinking Modern Prostheses*, p. 5.
7 Cara Kiernan Fallon, 'Walking cane style and medicalised mobility' in Elizabeth Guffey and Bess Williamson (eds), *Making Disability Modern: Design Histories* (London: Bloomsbury, 2020), p. 54.
8 Withey, *Self-Fashioning and Politeness*, pp. 109–110.
9 Guffey and Williamson (eds), *Making Disability Modern*, pp. 1–13.

10 Waters, 'Fashion in undress', p. 35; Susan Hiner, *Accessories to Modernity: Fashion and the Feminine in Nineteenth-century France* (Philadelphia: University of Pennsylvania Press, 2010), pp. 1–4; Ariel Beaujot, *Victorian Fashion Accessories* (London: Bloomsbury, 2012), p. 4.
11 Fallon, 'Walking cane style', pp. 44, 51.
12 Withey, *Self-Fashioning and Politeness*, p. 77; Jessica P. Clark, *The Business of Beauty: Gender and the Body in Modern London* (London: Bloomsbury Visual Arts, 2020), p. 6; Sadiah Qureshi, *People on Parade: Exhibitions and Anthropology in C19th Britain* (Chicago, IL: University of Chicago Press, 2011), 38–46.
13 *Oxford English Dictionary* (online). Fashion, *v*. 1. a. *transitive*. To give fashion or shape to; to form, mould, shape (either a material or immaterial object).
14 Lynda Nead, 'The layering of pleasure: women, fashionable dress and visual culture in the mid-nineteenth century', *Nineteenth-Century Contexts*, 35:5 (2013), 489–509 (p. 492).
15 Chris Otter, *The Victorian Eye: A Political History of Light and Vision in Britain, 1800–1910* (Chicago, IL, London: Chicago University Press, 2008), pp. 56–58; Martin Willis, *Vision, Science and Literature, 1870–1920: Ocular Horizons* (London: Pickering & Chatto, 2011), p. 198.
16 Julie Anderson, 'Separating the surgical and commercial: space, prosthetics and the First World War', in Jones (ed.) *Rethinking Modern Prostheses*, p. 174.
17 William White Cooper, *On Near Sight, Aged Sight, Impaired Vision and the Means of Assisting Sight*, 2nd edn (London: John Churchill, 1853), p. 6.
18 Coreen McGuire, 'Inventing amplified telephony: the co-creation of aural technology and disability' in Jones (ed.) *Rethinking Modern Prostheses*, p. 84.
19 Cooper, *On Near Sight*, p. 7.
20 Ibid., p. 5.
21 William Mackenzie, *A Practical Treatise on the Diseases of the Eye*, 4th edn (London: A. and G.A. Spottiswoode, 1854), p. 986.
22 Jaipreet Virdi, 'Medicalising deafness in Victorian London: the Royal Ear Hospital, 1816–1900', in Iain Hutchison, Martin Atherton and Jaipreet Virdi (eds), *Disability and the Victorians: Attitudes, Interventions, Legacies* (Manchester: Manchester University Press, 2020), p. 81.

23 Francisque Sarcey, *Mind your eyes! Advice to the Short-sighted, by their Fellow Sufferers*, trans. R.E. Dudgeon (London: Baillier, Tindall & Cox, 1886), pp. 3–5.
24 Richard Hengist Horne, 'Eyes and eyeglasses: a friendly treatise', *Fraser's Magazine* (December 1876), p. 698.
25 'Table talk', *Once a Week* (20 May 1871), p. 504.
26 See, for example, Mackenzie, *A Practical Treatise*, pp. 974–975.
27 John Zachariah Laurence, *Optical Defects of the Eye and their Consequences, Asthenopia and Strabismus* (London: Robert Hardewicke, 1865), pp. 69–70.
28 Ibid., p. 93.
29 Christopher Smith Fenner, *Vision: Its Optical Defects, and the Adaption of Spectacles* (London: Lindsay & Blakiston, 1875), p. 209.
30 See, for example, Carlisle Archive Centre, John Potter Dowell, DB9/4 Cash day book (sales) 1888–1890, 1 February 1888, 15 Febuary 1888, 23 February 1888, 6 March 1888, 6 April 1888, 9 September 1888, 10 October 1888, 20 November 1888, 13 December 1888, 20 December 1888, 2 January 1889, 11 January 1889, 14 January 1889, 23 January 1889, 29 January 1889, 30 January 1889.
31 Cooper, *On Near Sight*, p. 73.
32 Robert Brudenell Carter, *Eyesight: Good and Bad: A Treatise on the Exercise and Preservation of Vision*, 2nd edn (London: Macmillan, 1880), pp. 143–145. (NB Page numbers are different for American editions).
33 Simeon Snell, *Eye-strain as a Cause of Headache and Neuroses* (London: Simpkin, Marshall, Hamilton, Kent & Co., 1904); Sydney Stephenson, *Eye Strain in Everyday Practice* (London: The Ophthalmoscope Press, 1913).
34 Stephenson, *Eye Strain*.
35 Ibid., p. 1.
36 Ibid., pp. 27–33.
37 London Metropolitan Archive, A/KE/B/01/04/004, Royal London Ophthalmic Hospitals, complaint that there was little profit in supplying spectacles, prescription dated 1 November 1889.
38 Ibid., letter dated 29 November 1898.
39 Ibid., letter dated 12 May 1899.
40 See, for example, Peter John Brownlee, *The Commerce of Vision: Optical Culture and Perception in Antebellum America* (Philadelphia: University of Pennsylvania Press, 2019), pp. 10–11 and also p. 31 for examples of vision affecting people's livelihood in America.

41 London Metropolitan Archives, A/KE/B/01/04/004, Royal London Ophthalmic Hospitals, complaint that there was little profit in supplying spectacles, letter dated 23 May 1899.
42 Ibid., letter dated 20 May 1899.
43 Brownlee in *The Commerce of Vision* also recognised this in an American context: p. 65.
44 London Metropolitan Archives, A/KE/B/01/04/004, Royal London Ophthalmic Hospitals, complaint that there was little profit in supplying spectacles, letter dated 20 November 1900.
45 Ibid., letter dated 8 June 1899.
46 Ibid., letter dated 29 November 1898.
47 Dr Dyce Davidson, 'Professor Davidson on the eyesight: the evils of our school system', *Aberdeen Journal* (18 February 1885).
48 'Spectacles: to the editor of the *York Herald*', *The York Herald and General Advertiser* (21 October 1843), p. 8.
49 See, for example, *The Morning Post's* reflections on Robert Brudenell Carter's recent work (12 February 1880), p. 3.
50 'Some social changes in fifty years', *The Nineteenth Century: A Monthly Review* (March 1892), p. 465.
51 Charles Rolleston, 'Victims of industry', *Westminster Review* (January 1894), pp. 424–427.
52 London Metropolitan Archives, A/KE/B/01/04/004, Royal London Ophthalmic Hospitals, complaint that there was little profit in supplying spectacles, letters dated 29 November 1898, 22 May 1899, 23 May 1899, 18 May 1900 and printed in 'A petition on the subject of the use of appropriate lenses in certain diseases of the eye'.
53 See, in particular, Cooper, *On Near Sight*; Laurence, *Optical Defects of the Eye*; Charles Bell Taylor, *How to Select Spectacles in Cases of Long, Short, and Weak Sight*, 2nd edn (London: Cassell, 1889); A. Fournet, *Medical Spectacles and the Royal London Ophthalmic Hospital: Bloomfield Street, Moorfields, E.C.* (London: A. Fournet, 1894).
54 A similar diversity has been found in the American context: Brownlee, *Commerce of Vision*, p. 41.
55 'Talking and reading', *Saturday Review of Politics, Literature, Science and Art* (10 May 1862), p. 525.
56 Withey, *Self-Fashioning and Politeness*, p. 96.
57 Ryan Sweet, *Prosthetic Body Parts in Nineteenth Century Literature and Culture* (London: Palgrave Macmillan, 2022), pp. 227–272. Open access: DOI: 10.1007/978-3-030-78589-5.
58 Kay Heath, *Aging by the Book: The Emergence of Midlife in Victorian Britain* (New York: State University of New York Press, 2009).

59 Anne-Julia Zwierlein, Katharina Boehm and Anna Farkas (eds), *Interdisciplinary Perspectives on Aging in Nineteenth-Century Culture* (New York: Routledge, 2014), p. 1.
60 Heath, *Aging by the Book*, p. 16, and Ch. 5.
61 Sweet, *Prosthetic Body Parts*, pp. 229–234.
62 Ibid., p. 229.
63 Thos Harris & Son, *A Brief Treatise on the Eyes, Defects of Vision, and the Means of Remedying the Same by the Use of Proper Spectacles, Also Rules for judging when Spectacles are necessary, and Directions for selecting them* (London: Onwhyn, 1839), p. 17.
64 See, for example, 'Our London letter', *Leicester Chronicle and the Leicestershire Mercury* (2 July 1892); 'Notes from London', *The Sheffield & Rotherham Independent* (2 July 1892).
65 The National Archives, COPY 1/397/35, 6 July 1889.
66 The National Archives, COPY 1/19/110, 13 July 1872; COPY 1/366/247, 26 November 1883; COPY 1/414/928, 23 December 1893.
67 Heath, *Aging by the Book*, pp. 4–5; James Stark, *The Cult of Youth: Anti-Ageing in Modern Britain* (Cambridge: Cambridge University Press, 2020), pp. 6–7.
68 Heath, *Aging by the Book*, p. 16.
69 Hyung Wook Park, *Old Age, New Science: Gerontologists and Their Biosocial Visions, 1900–1960* (Pittsburgh, PA: University of Pittsburgh Press, 2016), pp. 26–7.
70 Sweet, *Prosthetic Body Parts*, pp. 229–230.
71 Pat Thane, *Old Age in English History: Past Experiences, Present Issues* (Oxford: Oxford University Press, 2002); Karen Chase, *The Victorians and Old Age* (Oxford: Oxford University Press, 2009), p. 6.
72 'Spectacles and chess from the *Sporting and Dramatic News*', The *Hull Packet and East Riding Times* (18 June 1880).
73 Melissa Dickson, 'Hats, cloaks and stethoscopes: the symbolic fashions of the nineteenth century medical practitioner', in Janine Hatter and Nickianne Moody (eds), *Fashion and Material Culture in Victorian Fiction and Periodicals* (Brighton: Edward Everett Root, 2019), pp. 105–120 (p. 105).
74 'Spectacles and chess', an extract from the *Sporting and Dramatic News* in *The Hull Packet and East Riding Times* (18 June 1880).
75 Ryan Sweet, '"Get the best article in the market": prostheses for women in nineteenth-century literature and commerce', in Jones (ed.) *Rethinking Modern Prostheses*, p. 117; Heath, *Aging by the Book*, p. 10.

76 Horne, 'Eyes and eyeglasses', pp. 698–722.
77 Fournet, *Medical Spectacles*, p. 21.
78 Horne, 'Eyes and eyeglasses', p. 714.
79 F.C. Donders, *On the Anomalies of Accommodation and Refraction of the Eye, with a Preliminary Essay on Physiological Dioptrics*, trans. William Daniel Moore (London: New Sydenham Society, 1864), p. 190.
80 Clark, *The Business of Beauty*, p. 4; Sweet, *Prosthetic Body Parts*, pp. 37–72.
81 'Dr Bell Taylor on school myopia or short sight, induced by over-education', *Nottinghamshire Guardian* (29 April 1887).
82 Heath, *Aging by the Book*, p. 12; Sweet, *Prosthetic Body Parts*, pp. 178–179.
83 Sweet, *Prosthetic Body Parts*, pp. 129–30.
84 'Our eyes and our industries', *Freeman's Journal and Daily Commercial Advertiser* (19 March 1887).
85 Steve Sturdy, 'The industrial body', in Roger Cooter and John Pickstone (eds), *Companion to Medicine in the Twentieth Century* (London: Routledge, 2003); Sarah F. Rose, *No Right to be Idle: the Invention of Disability, 1840s–1930s* (Chapel Hill: The University of North Carolina Press, 2017); Cindy LaCom, '"The time is sick and out of joint": physical disability in Victorian England', PMLA, 120:2 (2005), 547–552 (p. 547).
86 'Ladies' corner', *The Bristol Mercury* (7 February 1890); 'Fashion & household. Look to your eyes', *The Blackburn Standard* (15 February 1890); for medical texts, see for example, Snell, *Eye-Strain*, pp. 31–35; Stephenson, *Eye Strain*, pp. 1, 109–112.
87 Withey, *Self-Fashioning and Politeness*, pp. 109–110; Fallon, 'Walking cane style', p. 44.
88 See, for example, G.G. Bussey, patent number 3031 (20 February 1889); F. Merriman, patent number 4470 (14 March 1889); H. Supthut, patent number 20,623 (14 October 1889).
89 London Science Museum's Optics collection, object number 1931-789.
90 London Science Museum's Optics collection, and object numbers 1921-323/373–376; see, for example, Thos Harris & Son, *Daily News* (10 September 1877) and 'Francis', *The Wrexham Advertiser and North Wales News* (26 March 1887).
91 Jones, *Rethinking Modern Prostheses*, p. 10; Sweet, *Prosthetic Body Parts*, various; Stephen Mihm, 'A limb which shall be presentable in polite society: prosthetic technologies in the nineteenth century' in Katherine Ott, David Serlin and Stephen Mihm (eds), *Artificial Parts*

and *Practical Lives: Modern Histories of Prosthetics* (New York: NYU Press, 2002), throughout.
92 'Physiology of vision', *The London Journal and weekly record of literature, science and general information* (29 January 1848), p. 351.
93 W.M. Thackeray, 'The history of Pendennis; his fortunes, misfortunes, his friends and his greatest enemy', *The Athenaeum* (4 November 1848), pp. 1099–1101.
94 Peter Gurney, '"The age of veneer": Charles Dickens and the antinomies of Victorian consumer culture', *Dickens Quarterly*, 32:3 (2015), 229–246; Peter Gurney, *Wanting and Having: Popular Politics and Liberal Consumerism in England, 1830–1870* (Manchester: Manchester University Press, 2015).
95 Mihm, 'A limb which shall be presentable in polite society'.
96 'Over the teacups', *The Academy* (10 January 1891), p. 33.
97 Clark, *The Business of Beauty*, p. 7.
98 See, for example, *Arnold's Magazine of the Fine Arts, and Journal of Literature and Science* (May 1834) p. 10; *Examiner* (8 October 1837) p. 654, (19 November 1837) p. 750, (26 November 1837) p. 766, (3 December 1837) p. 783, (4 February 1838) p. 78, (25 February 1838) p. 126, (18 March 1838) p. 174; *Lloyd's Weekly London Newspaper* (26 September 1847), (24 October 1847); *Ipswich Journal* (3 June 1854); *Gentleman's Magazine* (April 1888) pp. 23–24, (March 1888) pp. 317–318, (June 1889) p. 622.
99 *Liverpool Mercury*, regularly between 10 March 1864 and 6 August 1864; *Ipswich Journal* (3 June 1854).
100 Neil Handley, *Cult Eyewear: The World's Enduring Classics* (London: Merrell Publishers Ltd, 2011).
101 'On spectacles and weak nerves', *Dublin University Magazine* (June 1877), pp. 781–782.
102 See, for example, 'The Criminal Amendment Act', *Reynolds' Newspaper* (8 June 1873).
103 For comparable accessories and/or clothing displayed in this way, see Hiner, *Accessories to Modernity*; Beaujot, *Victorian Fashion Accessories* and Breward, 'Masculine pleasures'.
104 Otter, *The Victorian Eye*, pp. 56–58; Willis, *Vision, Science and Literature*, p. 198.
105 See extract from Francis West's *Treatise on the Human Eye* in the *Operative* (27 January 1839); this is also commented on later in the century, in John Phillips, *Ophthalmic Surgery and Treatment: With Advice on the Use and Abuse of Spectacles* (London: W.B. Keen &

Co., 1869), p. 23; 'Health and Education Department', *Leeds Mercury* (5 October 1883); 'Health, beauty and the toilet', *Bow Bells: A magazine of general literature and art for family reading* (18 November 1892), p. 490; 'The eyesight of children', *Saturday Review of Politics, Literature, Science and Art* (18 July 1896), pp. 57–58.

106 Christopher Smith Fenner, *Vision: Its Optical Defects, and the Adaption of Spectacles* (London: Lindsay & Blakiston, 1875), pp. 220–222; Stephenson, *Eye Strain*, p. 1. See also 'Leicester School Board – the eyesight of school children', *Leicester Chronicle and the Leicestershire Mercury* (11 November, 1899).

107 Donders, *On the Accommodation and Refraction of the Eye*, p. 154.

108 Brownlee, *Commerce of Vision*, p. 72.

109 'Local chit-chat', *The Preston Guardian* (30 August 1884).

110 Jennie Batchelor and Cora Kaplan (eds), *Women and Material Culture, 1660–1830* (Basingstoke: Palgrave Macmillan, 2007); Nickianne Moody, 'Introduction' in Hatter and Moody (eds), *Fashion and Material Culture*, p. xvii.

111 'Spectacles', *Saturday Review of Politics, Literature, Science and Art* (21 August 1880), pp. 234–235.

112 Hiner, *Accessories to Modernity*, p. 3.

113 For the importance of modesty see Sweet, 'Get the best article in the market', p. 125.

114 'Boston society', *The Orchestra* (12 July 1872), p. 320.

115 'Odds and ends', *The Illustrated Police News* (25 April 1885).

116 'Quips and cranks', *The North-Eastern Daily Gazette* (25 July 1894); 'Scraps', *Aberdeen Weekly Journal* (10 April 1895).

117 Daryl Ogden, *The Language of the Eyes: Science, Sexuality, and Female Vision in English Literature and Culture, 1690–1927* (New York: State University New York Press, 2005), pp. 133–154.

118 Nead, 'The layering of pleasure', pp. 500–502.

119 'Feminine amenities', *Saturday Review of Politics, Literature, Science and Arts* (5 December 1868), pp. 743–744.

120 Moody, 'Introduction', p. xxv.

121 'Fashions for November, 1823', *La Belle Assemblée: or Court and Fashionable Magazine* (November 1823), pp. 214–219.

122 'A man of standing', *Fun* (13 August 1879), p. 61.

123 Brent, *The Cut of His Coat*; Breward, 'Masculine pleasures'.

124 Breward, 'Masculine pleasures', p. 70.

125 'On spectacles and weak nerves, by the London hermit', *Dublin University Magazine*, June 1877, pp. 780–786.

126 'Spectacles. From the *Quarterly Review*', *Trewman's Exeter Flying Post or Plymouth and Cornish Advertiser* (1 August 1850); *The Quarterly Review* (June 1850), pp. 51–52.
127 'Social statistics', *The London Reader: of literature, science, art and general information* (20 December 1873), p. 187.
128 Brent, *The Cut of His Coat*; Breward, 'Masculine pleasures'; Gurney, 'The age of veneer'; Gurney, *Wanting and Having*.
129 Sweet, *Prosthetic Body Parts*, p. 179.
130 'Frames by Lionel Laurance', *The Optician* (31 March 1898), p. 30.
131 William Smith Baxter, *Facts About Eyesight: Eyestrain and Spectacles* (Leeds: J. Lonsdale, 1898), p. 8.
132 See, for example, Arthur S. Underwood, 'An artificial nose affixed without spectacles', *The Lancet* (2 May 1896), p. 1232 and the Thackray Medical Museum's collection of later examples of hearing aid spectacles, Collections Online | Home (archive.org) (accessed: 3 February 2022).
133 'Spectacles for cosmetic effect', *The Optician* (6 July 1893), p. 692.
134 For broader discussion of the 'success' of prostheses in relation to (in)visibility, see Jaipreet Virdi, 'Between cure and prosthesis: "good fit" in artificial eardrums' in Jones (ed.), *Rethinking Modern Prostheses*, p. 51.
135 Jeffrey A. Brune and Daniel J. Wilson (eds), *Disability and Passing: Blurring the Lines of Identity* (Philadelphia, PA: Temple University Press, 2013), pp. 4–5.

Conclusion

> The hats, neckties, or boots of certain people seem as much parts of their persons as their noses or their whiskers; but no artificial adjuncts of the human body are so apparently identical with its nature as spectacles.[1]

In opening their article on 'Spectacles' in August 1880, the *Saturday Review of Politics, Literature, Science and Art* introduced spectacles as an 'artificial adjunct' so intimately connected to its wearer that it had a symbiotic relationship with a person's appearance and expression. It continued to argue that:

> [w]e know men who seem to smile with their spectacles, to frown, to sneer, and even to eat with them. They are the most prominent features, so to speak, of their countenances, and we should miss them as much as we should miss their eyes or their ears. Indeed, it would almost seem indecent if they were to take them off. It never occurs to us for a moment that they were born without them, nor would it strike us as strange if we were to see a little spectacle face peeping out of their babies' cradle.

But there was a caveat. This intimacy between assistive device and person was only attained by a 'habitual, chronic and incurable' use. Such habitual use was, however, becoming a Victorian necessity for all 'but few civilised people of a certain age' who managed to avoid a need for assistance. The article outlines a very broad, generalised transformation in the demographic of spectacle wearers and visual aid design. The 'brave' users of 'formidable looking' and 'cumbrous' frames were a phenomenon from half a century before that had only left a small visible trace in the 'out-of-the-way village shops' of rural Britain. For the Victorians, spectacles had 'moved with the times'. But, set against these overarching changes, the *Saturday Review* depicts the intersectionality of Victorian spectacle wear. Victorian visual aid use was deeply personal and the attitudes towards, and the realities of, spectacle wear were not experienced evenly. In the *Saturday Review* spectacles were displayed publicly to give certain users an air of piety, respectability, intelligence, even impertinence. But its writer also critiqued some aspects of public display – for example, the lady wearing spectacles outside the company of close companions. For some Victorians, spectacles became an accessory that they could proudly wear in their portrait; for others, 'they snatch[ed] their glasses from their nose when discovered as rapidly as a monkey would do it for them, if he were to get a chance, at the Zoological Gardens'.

In this book I have outlined how many of the *Saturday Review*'s observations about alterations in spectacle wear, while not originating in the Victorian period, were galvanised by the Victorians to serve the interest of a growing, and diverse, group of visual aid wearers. It therefore adds to growing literature that considers the lived experiences of a range of bodily conditions during the Victorian period at a time of rapid expansion, even 'maturation' of industry, urbanisation and medical investigation.[2] Several major Victorian developments influenced the style and wear of spectacles: alterations in mass manufacture, in medical intervention and in retailing practice and debates over certification. An unprecedented demand for both near and far visual acuity in a physical and economic environment that placed an emphasis on the ability to see also set the groundwork for enhanced vision testing and the regulation of visual acuity in areas such as a person's work. While the *Saturday Review* makes a geographical distinction in the use and distribution of spectacles, the reality was a somewhat more complex picture. Old-fashioned and out-dated dispensing practices and visual aid styles could be found in both rural localities and major towns: in the hands of itinerant pedlars and as part of the stock of miscellaneous sundries in bustling bazaars. But it is the increasing marginalisation of this type of practice that this book has traced. A variety of historical actors involved in the prescribing and dispensing process reframed spectacles and eyeglasses as a more complex device that needed to be accurately suited to their prospective wearer by an 'expert'. While there were a more diverse number of retailers engaging in spectacle dispensing by the 1890s, these retailers were increasingly being forced to upskill or contend with pressing demands for regulation, certification and control. The central premise of this book is that this alteration was not simply a medical intervention. It was part of a much broader reflective change in response to the growing demand for enhanced vision in an increasingly visual, modern Victorian world. Spectacle users had access to an unprecedented range of popular literature in pamphlets, leaflets, newspapers and periodicals to equip themselves with the latest knowledge on eyecare and spectacle use; the optician emerged, almost triumphant, with a definable and appropriate skillset; and visual aids could not be reduced to a basic, functional, utilitarian device because they had a much longer history of display.

I have approached the study of vision's importance to the Victorian with trepidation. The historical study of the senses can reinforce the historical hierarchies of the senses. Why write another book focusing only on the eye and vision and not engaging with the multisensory? It has not been my intention to push an agenda that vision superseded every other sense in Victorian Britain. Instead, by focusing on how a single sense was perceived and measured at a particular time, I aim to fill a gap that has been identified in sensory history. To return to Mark M. Smith's argument outlined in the Introduction, historians 'need to attend more to a broader variety of sensory constituencies' and have 'greater sensitivity to the senses and disabilities'.[3] In this book I have attempted to make a start; the Victorian measurement of the eye shaped the attitudes to, expectations and understanding of the 'functional body'. My aim has been to demonstrate that there is a real opportunity to connect recent trends in disability history which have focused on the way in which disability has been historically, and is presently, classified with recent trends in sensory history that seek to reconceptualise sensory capacity as a varied and subtle continuum.[4] The specifically Victorian experience of seeing that I have uncovered was in many ways paradoxical. The Victorian environment identified and exposed a large number of 'unseeing' people: people who were not capable of meeting the visual acuity requirement necessary to function within a visual world of marketing, navigation, faster transport, enclosed spaces, small print and employments that demanded considerable ocular labour through either close or distant work. But it was also an environment that had a far greater understanding of visual acuity – be it medical or popular – and emphasised the importance of ocular health and eye care. This is not to be environmentally determinist; the environment in and of itself did not shape perceptions of visual acuity, but the variety of historical actors that capitalised upon it did.

Scholars have argued that, to unpick a fuller understanding of sensory experience in the past, we need to contextualise the sensory environment and understand the historical specificity of people's bodies.[5] The Victorian experience of seeing was different from that of today and it is from this context that I argue that the Victorian use and marketing of spectacles offers a new perspective on the way in which technology and its design shapes our understanding of the functional and non-functional body. Spectacles – and diagnostic technology in the vision testing process – defined the categories, limits and parameters of seeing. In influencing contemporary understandings of the human body's visual capacity, spectacles were just as transformative. Victorians who had described their visual state as 'blind' experienced first-hand what it was like to be able to 'see'. This reconceptualisation of a sensory threshold had a profound impact on people's sensory experience and the way they categorised their own bodily state. It had many positives. Indeed, it is easy to read Victorian insights into the 'restorative power' of spectacles and conclude, as in other scholarly work, that spectacles' capability of meeting the needs of its users was unique. But the reality was much more complex, and the idea of a 'cure' was vexed. Adapting optical lenses to a person's vision was not always successful or straightforward. Moreover, as I have shown, spectacle use and objective vision testing methods – the processes that determined the sensory thresholds – were governed by bureaucracy, medical intervention and the commercial interests of retailers. While marketed and discussed in medical texts as objective, these diagnostic tools were interpreted in subjective and contested ways.

Visual aids therefore offer an insight into not only how technology can alter our understanding of the body but also the historical and contemporary actors that controlled the way in which a person's sensory capacity was measured, categorised and, subsequently, should be felt. This was not an irrefutable, passive process, and the categorisation of vision was met with resistance both in the medical profession and in the broader conduct of spectacle users who did not regard or adopt spectacles as a medical device or agree with workplace vision testing standards.[6] Crucially, the number of vision 'defects' being diagnosed complicated categories of normalcy and abnormalcy and, coupled with the 'natural' requirement for spectacles in old age, the Victorians created a mainstream spectacle market. As Coreen McGuire has recently argued in her study of technology and the objective measurement of disability in the period between the First and Second World Wars, 'the heterogeneity and uniqueness of individual bodies has often been at odds with the standardisation of biomedicine'.[7] This was also the case for Victorian medical practitioners attempting to quantify 'normal' vision. Indeed, in this case, I can take the argument a step further. In considering the role of technology in assisting and 'treating' categories of abnormalcy, this book has revealed that Victorian spectacle wearers also moved in different environments which served to further cultivate unique, individual and varied responses to visual (in)capacity and visual aid use. The design and consumption of spectacles tackled a key question – how in control should we be of the way in which our sensory capacities are measured and perceived? – as well as offering technology as a means to subvert categories of disease or abnormalcy by adopting different styles of frames.

Chapters Two to Four reveal the power dynamics at play in the categorisation of visual acuity, particularly between medical practitioners and an emerging group of certificated opticians – the forerunners of the later professionally titled optometrists on the high street. The Victorians did not manage to medicalise vision testing and its treatment, and neither did they solve inter- and intra-professional tensions in vision testing, retail and dispensing. Caroline Weaver, for example, has tracked the increasing 'bureaucratisation of the senses' in the period between the World Wars as part of continued fears of blindness because of modern society.[8] She highlights that, like their 1890s contemporaries, ophthalmologists placed pressure on state intervention as part of their ambitions to position themselves as public health professionals and experts. In contrast, opticians faced restrictions and a legislative defeat in the Optical Practitioners Bill of 1927 – a bill to create a professional register – which put on hold the regularisation of their practice until the 1930s. Echoes of the 1890s conflict discussed in Chapter Four hung over optical practice in the first half of the twentieth century. The Victorians had not established battle lines; they had established a 'no man's land' of blurred expertise, one that inevitably led to conflict but also was a space for shared concerns to be raised. Victorian medical intervention in vision testing therefore needs to be seen as part of broader historical processes: ophthalmologists' continued need to establish their medical professional identity in a climate of mistrust for medical specialism, the general medical profession's approach to dealing with other nascent professions and the increasing pressure on late Victorian and early twentieth-century medical practitioners to engage with, and control, a growing commercial market for medical products. Ultimately, the Optician's Act in 1958 successfully established a professional register of optometrists. Continually – in the 1890s, the 1930s and in the dispensing of spectacles as part of the National Health Service (NHS) after the Second World War – the level of demand for optical care shaped opticians' eventual success. Medical practitioners repeatedly grappled with their attempts to shape simple refractive vision errors as a medical problem while falling short of the capacity to provide such care.

The premise of the NHS's initial provision of ophthalmic services in 1946, as part of the National Health Service Acts, was that eye care should be accessible to all.[9] As Chapter Three demonstrated, concurrent with the acknowledgement of spectacles as a transformative assistive aid was the idea that they were a 'basic right'. The 'modern' visual aid was capable of transforming a person's vision, but this 'modern' aid was also tackling the 'problem' of visual acuity that 'modernity' had brought to the fore and created. In studies of the historical norm, the 'normal' has often been conflated with the 'ideal' rather than the 'average'.[10] For the Victorian, 'normal' vision – or 20:20 vision – was conceived as an ideal because it was perceived as a *necessary* reality when attempting to navigate safely in a modernised physical environment and economy. The demand for visual aids, coupled with the idea that frames should be accessible, influenced spectacle design. But importantly, spectacle design also influenced spectacle use and how any variance in visual acuity would be perceived. Certain visual aid designs, although socially stratified and shaped within the parameters of fashion and stigma, had the capacity to be promoted as 'universal', or even 'normal'. It is helpful here to look ahead to the discussions of 'utility spectacles' in the mid-twentieth century as part of the NHS scheme. Jo Gooding in her study of NHS spectacles included ministerial notes from 1950, which argued 'we must... be careful not to admit a "fashion" element which might easily increase demand unnecessarily'.[11] The Ministry of Health's wariness of fashion in NHS spectacle design is a nod to the historical role of fashion and style in influencing the extent of spectacle use. A variety of designs since the Victorian period had ensured visual aids could be displayed or hidden in the workplace, at social events, at home and on the busy thoroughfares of towns and cities. Indeed, an outcome of twentieth-century reticence to request more stylish NHS spectacles was the rise of an external, private market, one that, like its Victorian counterparts, treated its products as a commercial commodity, or fashionable item to be displayed.

The historical trends of visibility and boldness in many aspects of spectacle design can be contrasted with the invisible styles of other, more ubiquitous assistive devices, such as the artificial hearing aid. Why is this the case? In this book I have argued that it is not just about the demand for visual aids. Frequency did play a part in the assimilation of visual aids as an assistive device, but far more profound was the adaptability of visual aid frames to suit individual bodies, individual circumstances and individual environments. Design and related technology advancements were at the heart of creating a fashionable, more normal, assistive device, a device that could form part of the identity of its wearer. More work, however, needs to be done to fully understand these processes. Comparative studies are required into the way in which the marketing of assistive devices has a bearing on the conceptualisation of disability. How do spectacles interact with other types of visual aid, or compare to assistive devices that serve to enhance the functionality of different parts of the body? The study of Victorian visual aids would also benefit from an analysis of visual and literary representations. Recent work on the literary representation of nineteenth-century prostheses, for example, has demonstrated the usefulness of this type of source for exploring popular opinions and the relative assimilation of a variety of devices, including artificial legs, dentures and wigs.[12] Work that connects discrete epochs of time that have been explored in the history of vision testing and spectacle use is also required to assess how these trends played out on a broader scale from the early twentieth century until the Opticians Act of 1958 and beyond into the present day.

Crucially, the historical study of spectacles, and the provision of vision testing, needs to be placed in a global context. This book has outlined, where relevant, the relationships and debates formulated between Britain, America and Europe in the *fin de siècle*, but it has not considered non-Western perspectives. Studies of colonial experiences of, for example, blindness emphasise a distinction and tension between native ideals and methods, and imported, Western technologies and ideas.[13] Such work is vital if we consider the *World Report on Vision* of the World Health Organization (WHO), which was published in 2019. It estimated 123.7 million people have a vision impairment from unaddressed refractive vision error and 826 million live with a vision impairment from unaddressed presbyopia.[14] In its introductory section it argued that 'the global need for eye care is projected to increase dramatically in the coming decades'.[15] This crisis, however, is felt unevenly and is affected by geographic disparities in economic circumstances, access to appropriate health care and relative engagement in close work activity. Spectacles have also been included, since 2018, on the WHO's Priority Assistive Products List as an 'essential item' that is not yet part of universal health coverage.[16] In many ways, the WHO's report and actions would resonate with Victorians and their fears and concerns within both their nation-states and their colonies. Associated twenty-first-century media content draws upon rhetoric similar to that published and promoted by Victorian medical practitioners and writers in the Victorian popular press – although concerns about newspaper reading have been replaced by fear of digital screens. Such parallels serve to reinforce the way in which our environment, and global context, mediates our experience of sensing and the role of technology in seeking to mitigate or accentuate its effects. It invites us to consider the deeply contextual experience of 'seeing' which, during and since the Victorian period, has been shaped by technology and the design and adoption of the visual aid.

The introduction of technology, and the adaptation of existing technologies, altered the thresholds and categorisation of Victorian vision, whether through demanding a certain level of functionality, shaping the diagnostic process, or through the way people responded to assistive devices seeking to enhance and restore. I argue that, at a time when vision classifications increasingly sought to homogenise the categories of visual capacity, visual aid design and functionality was positioned better than ever to suit and adapt to the individual and specific needs of its wearer. In this study I have sought to emphasise the individuality of spectacle wear within the broader, more sweeping, narratives of medicalisation and normalisation. This captures both the ways in which the concept of normative vision – and attempts to medicalise vision – did have influence, and those in which it was contested. 'Normative vision' did as much to expose the individuality of a person's visual capacity as it did to classify different types of visual capacity. Here, I argue that a multisensorial way of approaching the past is key to understanding the lived experiences and historical variability of using an everyday object. 'Sensing' the past through touch – exploring the traces of those not in the written record – has been vital for uncovering the covert ways individuals resisted medical classifications of vision, spectacles and eyeglasses. The very materiality of visual aids – the design choices, materials used and evidence of wear, tear and repair – offers an insight into the way in which an assistive device was used and dispensed in ways outside medical control, as well as the emotional attachment(s) of former wearers. Walking into storerooms of objects is a very tangible encounter with the Victorian experience of seeing. Their vision was mediated through the lenses in medical practitioners and opticians' trial cases, the ophthalmoscopes that observed their eyes and an even greater variety of frame styles. A history of spectacles would be too narrowly framed without analysing the materiality and design of its associated objects, whether that be in the diagnostic technologies to subjectively measure visual capacity or in the assistive aids that allowed its users to simultaneously display, mask or enhance their functional visual capacity or physical appearance. As the design and response to Victorian spectacles and eyeglasses reveal, 'normal' vision was often at odds with the increasingly mainstream visual aid. Concurrently, visual aids alleviated the symptoms of sight loss and in turn recategorised blindness as a totalising state of sight loss incapable of amelioration through technological means.

Notes

1 'Spectacles', *Saturday Review of Politics, Literature, Science and Art* (21 August 1880), pp. 234–235.
2 Iain Hutchison, Martin Atherton and Jaipreet Virdi, 'Introduction', in Iain Hutchison, Martin Atherton and Jaipreet Virdi (eds), *Disability and the Victorians: Attitudes, Interventions, Legacies* (Manchester: Manchester University Press, 2020), p. 14.
3 Mark M. Smith, *A Sensory History Manifesto* (Philadelphia: Pennsylvania State University Press, 2021), p. 65.
4 Smith, *A Sensory History Manifesto*, pp. 29–31, 79; see also Jonathan Reinarz, *Past Scents: Historical Perspectives on Smell* (Urbana: University of Illinois Press, 2014); Alain Corbin, *Time Design and Horror: Toward a History of the Senses*, trans. by Jean Birrell (Cambridge: Polity Press, 1995); Alain Corbin, *The Foul and The Fragrant: Odor and the French Social Imagination* (Cambridge, MA: Harvard University Press, 1996).
5 Smith, *A Sensory History Manifesto*, p. 89; Joy Parr, *Sensing Changes: Technologies, Environments and the Everyday, 1953–2003* (Vancouver: UBC Press, 2010), pp. 189–191.
6 See, for example, Jordanna Bailkin, 'Colour problems: work, pathology and perception in modern Britain', *International Labour and Working-Class History*, 68 (2005), 93–111.
7 Coreen McGuire, *Measuring Difference, Numbering Normal: Setting the Standards for Disability in the Interwar Period* (Manchester: Manchester University Press, 2020), p. 207.
8 Caroline Weaver, 'Eyesight and governance in Britain: bureaucracy and the senses in the 1920s', *Social History of Medicine*, 27:2 (2014), 241–259.

9 Jo Gooding, 'Rather unspectacular: design choices in National Health Service glasses', *Science Museum Group Journal*, 7 (2017). DOI: 10.15180/17070.
10 See Chapter Two.
11 Gooding, 'Rather unspectacular'.
12 Ryan Sweet, *Prosthetic Body Parts in Nineteenth Century Literature and Culture* (London: Palgrave Macmillan, 2022). Open access: DOI: 10.1007/978-3-030-78589-5.
13 Aparna Nair, '"They shall see his face": blindness in British India, 1850–1950', *Medical History*, 61:2 (2017), 181–199; Aparna Nair, 'Of ear trumpers, audiphones, and the language of fingers (*kar pallavi bhasha*): technologies for the deaf in British India', in Elizabeth Guffey and Bess Williamson (eds), *Making Disability Modern: Design Histories* (London: Bloomsbury, 2020).
14 World Health Organization, *World Report on Vision* (Geneva: World Health Organization, 2019), pp. 26–27.
15 Ibid., p. x.
16 Priority Assistive Products List (APL) (who.int) (accessed 15 February 2022).

Bibliography

Primary Sources

Science Museum Object Collections

Art collection
Ophthalmology collection
Optics collection

Archives

Boots Archives, Nottingham
Cambridgeshire Archives
Cambridge University Library
Carlisle Archive Centre
College of Optometrists
Devon Heritage Centre
London Metropolitan Archives
London School of Economics
Science Museum's Library and Collections
Sheffield City Archives
Somerset Heritage Centre
The National Archives
Thackray Museum, Leeds
Trinity College Library, Cambridge
University of Leicester, Special Collections Online
University of Nottingham Archives Manuscripts & Special Collections
Wellcome Library
West Sussex Record Office

British Library 19th Century Newspaper Archive

Aberdeen Journal
The Belfast News-Letter
Birmingham Daily Post
The Blackburn Standard
The Bradford Observer
Bristol Mercury
The Bury and Norwich Post and Suffolk Herald
Caledonian Mercury
Cheshire Observer and Chester, Birkenhead, Crewe and North Wales Times
Daily News
Derby Mercury
The Dundee Courier & Argus
The Essex Standard and General Advertiser for the Eastern Counties
Freeman's Journal and Daily Commercial Advertiser
Glasgow Herald
Hampshire Advertiser & Salisbury Guardian
The Huddersfield Daily Chronicle
Ipswich Journal
The Isle of Man Times and General Advertiser
Isle of Wight Observer
Jackson's Oxford Journal
Leeds Mercury
Leicester Chronicle and the Leicestershire Mercury
Lloyd's Weekly London Newspaper
Manchester Times
The Morning Post
The Newcastle Weekly Courant
North-Eastern Daily Gazette
North Wales Chronicle
Northern Echo
Nottinghamshire Guardian
Operative
The Pall Mall Gazette
Preston Chronicle
The Preston Guardian
Reynold's Newspaper
The Royal Cornwall Gazette Falmouth Packet, Cornish Weekly News, & General Advertiser
The Sheffield & Rotherham Independent

Southampton Herald
The Standard
The Star
Trewman's Exeter Flying Post or Plymouth and Cornish Advertiser
The Wrexham Advertiser, and North Wales News
The Yorkshire Herald

British Periodicals Collection I and II

The Academy
Arnold's Magazine of the Fine Arts, and Journal of Literature and Science
The Athenaeum
La Belle Assemblée: or Court and Fashionable Magazine
Bentley's Miscellany
Blackwood's Edinburgh Magazine
Book-Lore
Bow Bells: A Magazine of General Literature and Art for Family Reading
Bradshaw's Manchester Journal
Chambers's Edinburgh Journal
Chambers's Journal of Popular Literature, Science and Arts
The Critic
Dublin University Magazine
The Edinburgh Review
Examiner
Fraser's Magazine
Fun
Gentleman's Magazine
The Graphic
The Leisure Hour
The Literary Gazette: A Weekly Journal of Literature, Science, and the Fine Arts
The London Dispatch and People's Political and Social Reformer
The London Journal and weekly record of literature, science and general information
The London Reader: of Literature, Science, Art and General Information
Longman's Magazine
Monthly Magazine, or British Register
The Musical World
The Nineteenth Century: A Monthly Review
The North British Review
Once a Week
The Orchestra

The Practical Teacher
The Saturday Magazine
Saturday Review of Politics, Literature, Science and Art
Tait's Edinburgh Magazine
Time
Westminster Review

Professional Journals

British Journal of Ophthalmology
British Medical Journal
The Optician

Published Works Cited in the Text

Ackland, William, *Hints on Spectacles: When to Wear and How to Select Them* (London: Horne & Thornthwaite, 1866).

Arlidge, J.T., *The Hygiene Diseases and Mortality of Occupations* (London: Perceval, 1892).

Baxter, William Smith, *Facts About Eyesight: Eyestrain and Spectacles* (Leeds: J. Lonsdale, 1898).

Beer, Georg, *The Art of Preserving the Sight Unimpaired to an Extreme Old Age; and of Re-establishing and Strengthening it When it is Become Weak* (London: Henry Colborn, 1815).

Berry, George A., *Diseases of the Eye: A Practical Treatise for Students of Ophthalmology*, 1st edn (Edinburgh and London: Young J. Pentland, 1889).

Bickerton, Thomas H., *Colour Blindness and Defective Eyesight in Officers and Sailors of the Mercantile Marine: A Criticism of the Board of Trade Tests* (Edinburgh: James Thin, 1890).

Black, George, *Eyesight and How to Care for it* (London: Ward, Lock, 1888).

Bowman, William, 'An address delivered at the opening of the section of Ophthalmology', *British Medical Journal* (13 August 1881), 277–279.

Browning, John, *Our Eyes and How to Preserve Them from Infancy to Old Age. With Special Information About Spectacles*, 7th edn (London: Chatto & Windus, 1887).

Carter, Robert Brudenell, *Eyesight: Good and Bad: A Treatise on the Exercise and Preservation of Vision*, 2nd edn (London: Macmillan, 1880).

Carter, Robert Brudenell, *Eyesight in Schools: A Paper Read Before the Association of Medical Officers of Schools on April 15th, 1885* (London: Harrison and Sons, 1885).

Bibliography 261

Cooper, William White, *On Near Sight, Aged Sight, Impaired Vision and the Means of Assisting Sight*, 2nd edn (London: John Churchill, 1853).

Donders, F.C., *On the Anomalies of Accommodation and Refraction of the Eye, with a Preliminary Essay on Physiological Dioptrics*, trans. William Daniel Moore (London: New Sydenham Society, 1864).

Dunn, Percy, 'A model eye', *British Medical Journal* (10 January 1903).

Emerson, William, *The Elements of Optics. In Four Books* (London: J. Nourse, 1768).

Farquharson, Robert, *School Hygiene and Diseases Incidental to School Life* (London: Smith, Elder, 1885).

Fenner, Christopher Smith, *Vision: Its Optical Defects, and the Adaption of Spectacles* (London: Lindsay & Blakiston, 1875).

Fournet, A., *Medical Spectacles and the Royal London Ophthalmic Hospital: Bloomfield Street, Moorfields, E.C.* (London: A. Fournet, 1894).

Grimshaw, John, *Eyestrain and Eyesight; How to Help the Eye and Save the Sight* (London: J. & A. Churchill, 1907).

Harris, Joseph, *A Treatise of Optics: Containing Elements of the Science, In Two Books* (London, 1775).

Harris, Thos, & Son, *A Brief Treatise on the Eyes, Defects of Vision, and the Means of Remedying the Same by the Use of Proper Spectacles, Also Rules for judging when Spectacles are necessary, and Directions for selecting them* (London: Onwhyn, 1839).

Hartridge, Gustavus, *The Refraction of the Eye: A Manual for Students* (London: J & A Churchill, 1884).

Horne, Richard Hengist, 'Eyes and eyeglasses: a friendly treatise', *Fraser's Magazine* (December 1876), 698–722.

Horner, Johann Friedrich, *On Spectacles: Their History and Uses* (London: Balliere, Tindall & Cox, 1887).

Hudson, J.T., *Spectalaenia; or the Sight Restored, Assisted and Preserved by the Use of Spectacles* (London: Simpkin and Marshall, 1833).

Hudson, J.T., *Useful Remarks upon Spectacles, Lenses, and Opera-Glasses; with Hints to Spectacle Wearers and others; being an epitome of practical and useful knowledge upon this popular and important subject* (London: Joseph Thomas, 1840).

James, Robert, *A Medicinal Dictionary, including physic, surgery, anatomy, chymistry and botany*, vol. III (London: T. Osborne, 1743–5).

Juler, Henry, 'On the best methods of diagnosing and correcting the errors of refraction', *British Medical Journal* (27 December 1884), 1274–1275.

Johnson, George Lindsay, 'The development of optics in the present century', *The Optician* (21 April 1898).

Keith, J. Gray, *Facts Relating to Spectacles: How, When, Where and Why to Wear Them, with Rules for the Preservation of the Sight, And*

Cautions as to how thousands Ruin their Sight (Glasgow: David Bryce & Son, 1889).

Kitchiner, W., *The Economy of the Eyes: Precepts for the Improvement and Preservation of Sight* (London: Hurst Robinson & Co., 1824).

Laurance, Henry, *The Eye in Health and Disease: With Hints on the Choice and Use of Spectacles*, 3rd edn (London: Love Brothers, 1888).

Laurence, John Zachariah, *Optical Defects of the Eye and their Consequences, Asthenopia and Strabismus* (London: Robert Hardewicke, 1865).

Lewis, William, *Medical Essays and Observations, published by a society in Edinburgh, containing meteorology, mineral waters, material medica and pharmacy*, vol. I (London, 1746).

Liebreich, Richard, *School Life in Its Influence on Sight and Figure: Two Lectures* (London: J. & A. Churchill, 1878).

Long, Charles A., *Spectacles: When to Wear and How to Use Them: Addressed to Those Who Value Their Sight* (London: Bland and Long, 1855).

Longmore, T., *The Optical Manual: or, Handbook of Instructions for the Guidance of Surgeons in Testing the Range and Quality of Vision of Recruits and Others Seeking Employment in the Military Services of Great Britain* (London: HMSO, 1885).

Loring, Edward G., *Is the Human Eye Changing its Form Under the Influence of Modern Education* (Publisher not identified, 1878).

Love, D., 'The vision of school children', *British Medical Journal* (25 March 1899), 763.

Mackenzie, William, *A Practical Treatise on the Diseases of the Eye*, 4th edn (London: A. and G.A. Spottiswoode, 1854).

Macleod, Kenneth, 'Remarks on the physical requirements of the public services', *British Medical Journal* (11 May 1895), 1021–1025.

Mayhew, Henry, *London Labour and the London Poor, A Cyclopaedia of the Condition and Earnings or Those that Will Work, Those that Cannot Work, and Those that Will Not Work*, vol. 1 (London: Griffin, Bohn and Company, 1851).

Middlemore, Richard, *Introductory lecture on the Anatomy, Physiology, and Diseases of the Eye, Delivered at the Birmingham Royal School of Medicine and Surgery, 4 October 1839* (London: S. Longman, Orme, Brown, Green and Longmans, 1839).

Middleton, Erasmus, *The New Complete Dictionary of Arts and Sciences*, vol. II (London, 1778).

Morton, A. Stanford, *Refraction of the Eye: It's Diagnosis and the Correction of its Errors: with a Chapter on Keratoscopy* (London: H.K. Lewis, 1881).

Motherby, George, *A New Medical Dictionary; Or, General Repository of Physic* (London: J. Johnson, 1791).

Bibliography 263

Newsholme, Arthur, *School Hygiene: The Laws of Health in Relation to School Life* (London: Swan Sonnenschein, Lowrey, 1887).

Phillips, John, *Ophthalmic Surgery and Treatment: With Advice on the Use and Abuse of Spectacles* (London: W.B. Keen & Co., 1869).

Phillips, R.J., *Spectacles and Eyeglasses: Their Forms, Mounting and Proper Adjustments*, 2nd edn (*The Optician and Photographic Trades Review*, 1900).

Roberts, C.D., *The Detection of Colour-Blindness & Imperfect Eyesight by the Methods of Dr Snellen, Dr Daae, and Prof Holmgren: with a table of coloured Berlin wools and sheet of test types* (London: David Bogue, 1881).

Roosa, D.B., *Defective Eyesight: The Principles of its Relief by Glasses* (London: Macmillan, 1899).

Ross, Andrew, *On the Use and Abuse of Spectacles* (London: R. Kinder, 1840).

Rowley, William, *A Treatise on One Hundred and Eighteen Principle Diseases of the Eyes and Eyelids* (London: J. Wingrave, 1790).

Rowley, William, *Dr Rowley's Rational Practice of Physic, in Four Volumes*, vol. III (London, 1793).

Sarcey, Francisque, *Mind your eyes! Advice to the Short-sighted, by their Fellow Sufferers*, trans. R.E. Dudgeon (London: Baillier, Tindall & Cox, 1886).

Smee, Alfred, *Vision in Health and Disease: the Value of Glasses for its Restoration and the Mischief Caused by their Abuse* (London: Horn, Thornthwaite and Wood, 1847).

Smith, Addison, *Visus Illustratus; or, the sight rendered clear and indistinct* (London, 1783).

Smith, Egerton, *Hints to the Wearers of Spectacles; or a Concise Statement of the Comparative Merits of Pebbles and Glasses, When Used as Spectacle Eyes* (Liverpool, 1819).

Smith, Priestley, *Short Sight in Relation to Education* (Birmingham: The Midland Educational Company, 1880).

Snell, Simeon, *Eye-strain as a Cause of Headache and Neuroses* (London: Simpkin, Marshall, Hamilton, Kent & Co., 1904).

Snell, Simeon, *Influences of School Life on Eyesight* (London: Wyman & Sons, 1884).

Snellen, H., *Test-Types for the Determination of the Acuteness of Vision* (1862).

Stephenson, Sydney, *Eye strain in Everyday Practice* (London: The Ophthalmoscope Press, 1913).

Taylor, Charles Bell, *How to Select Spectacles in Cases of Long, Short and Weak Sight*, 2nd edn (London: Cassell, 1889).

Taylor, Harry L., *The Practical Opticians Guide: An Elementary Course for Opticians* (Birmingham: J. & H. Taylor, 1897).

Warlomont, E., 'On the use of optometers for the examination of soldiers and workmen employed on the railroad', *British Medical Journal* (5 March 1881), 333–336.

Wells, Joel Soelberg, *On Long, Short and Weak Sight and Their Treatment by the Scientific Use of Spectacles* (London: J.A. Churchill, 1862).

West, Francis, *A Familiar Treatise on the Human Eye: Containing Practical Rules that will Enable all to Judge what Spectacles are Best Calculated to Preserve their Eyes to Extreme Old Age*, 2nd edn (London: W. Ackrill, 1827).

Willich, F.M., *Lectures on Diet and Regimen: Being a systematic inquiry into the most rational means of preserving health and prolonging life* (London: A. Strahan, 1800).

Secondary Sources

Abbott, Andrew, *The System of Professions: An Essay on the Division of Expert Labour* (Chicago: The University of Chicago Press, 1988).

Adams, Rachel, Benjamin Reiss and David Serlin (eds), *Keywords for Disability Studies* (New York: New York University Press, 2015).

Albert, Daniel M. and Diane D. Edwards (eds), *The History of Ophthalmology* (Oxford: Blackwell Science, 1996).

Alexander, David, *Retailing in England during the Industrial Revolution* (London: The Athlone Press, 1970).

Almond, Gemma, 'Vision testing in late nineteenth and early twentieth-century Britain: opticians, medical practitioners and the battle for professional authority', *Social History of Medicine*, 35:1 (2022), 237–258. DOI: 10.1093/shm/hkab122.

Andressen, Michael, *Spectacles: From Utility Article to Cult Object* (Stuttgart: Arnoldsche, 1998).

Arnold, Ken and Danielle Olsen (eds), *Medicine Man: The Forgotten Museum of Henry Wellcome* (London: The British Museum Press, 2011).

Atchison, David A. and W. Neil Charman, 'Thomas Young's contribution to visual optics: The Bakerian Lecture "On the mechanism of the eye"', *Journal of Vision*, 10:12 (2010), 1–16.

Bailkin, Jordanna, 'Colour problems: work, pathology and perception in modern Britain', *International Labour and Working-Class History*, 68 (2005), 93–111.

Barck, Carl, *The History of Spectacles, Originally Delivered as a Lecture Before the Academy of Science* (Reprinted from the Open Court for April, 1907).

Barker, Hannah, 'Medical advertising and trust in late Georgian England', *Urban History*, 36:3 (2009), 379–398.

Batchelor, Jennie and Cora Kaplan (eds), *Women and Material Culture, 1660–1830* (Basingstoke: Palgrave Macmillan, 2007).
Beaujot, Ariel, *Victorian Fashion Accessories* (London: Bloomsbury, 2012).
Benson, John, 'Drink, death and bankruptcy: retailing and respectability in late Victorian and Edwardian England', *Midland History*, 32:1 (2007), 128–140.
Berry, Helen, 'Polite consumption: shopping in eighteenth-century England', *Transactions of the Royal Historical Society*, 12 (2002), 375–394.
Birch, Susan and Michael Rembis (eds), *Disability Histories* (Chicago: University of Illinois Press, 2010).
Borsay, Anne and Billie Hunter (eds), *Nursing and Midwifery in Britain since 1700* (Basingstoke: Palgrave Macmillan, 2012).
Bourke, Joanna, *Dismembering the Male: Men's Bodies, Britain and the Great War* (London: Reaktion, 1999).
Brent, Shannon, *The Cut of His Coat: Men, Dress and Consumer Culture in Britain 1860–1914* (Athens, OH: Ohio University Press, 2006).
Breward, Christopher, 'Masculine pleasures: metropolitan identities and the commercial sites of dandyism, 1790–1840', *The London Journal*, 28:1 (2003), 60–72.
Briggs, Asa, *Victorian Things* (London: B.T. Batsford Ltd, 1998).
Bronstein, Jamie L., *Caught in the Machinery: Workplace Accidents and Injured Workers in Nineteenth-Century Britain* (Stanford, CA: Stanford University Press, 2008).
Brown, Michael, 'Medicine, reform and the "end" of charity in early nineteenth-century England', *English Historical Review*, 124:511 (2009), 1353–1388.
Brown, Michael, *Performing Medicine: Medical Culture and Identity in Provincial England, c.1760–1850* (Manchester: Manchester University Press, 2010).
Brownlee, Peter John, *The Commerce of Vision: Optical Culture and Perception in Antebellum America* (Philadelphia: University of Pennsylvania Press, 2019).
Brune, Jeffrey A. and Daniel J. Wilson (eds), *Disability and Passing: Blurring the Lines of Identity* (Philadelphia, PA: Temple University Press, 2013).
Burns, Arthur and Joanna Innes (eds), *Rethinking the Age of Reform: Britain 1780–1850* (Cambridge: Cambridge University Press, 2003).
Bynum, W.F. and Roy Porter (eds), *Medical Fringe & Medical Orthodoxy 1750–1850* (London: Croom Helm, 1987).
Canguilhem, Georges, *The Normal and the Pathological* (New York: Zone Books, 1991).
Carpenter, Mary, *Health, Medicine and Society in Victorian England* (Santa Barbara, CA: Praeger, 2009).

Casper, Stephen, *The Neurologists: A History of a Medical Specialty in Modern Britain, c.1789–2000* (Manchester: Manchester University Press, 2016).
Champness, Richard, *A Short History of the Worshipful Company of Spectacle Makers up to the Beginning of the Twentieth Century* (London: Apothecaries Hall, 1952).
Chase, Karen, *The Victorians and Old Age* (Oxford: Oxford University Press, 2009).
Clark, Jessica P., *The Business of Beauty: Gender and the Body in Modern London* (London: Bloomsbury Visual Arts, 2020).
Clifton, Gloria, *Directory of British Scientific Instrument Makers 1550–1851* (London: Zwemmer, 1995).
Cody, Lisa Forman, '"No cure, no money", or the invisible hand of quackery: the language of commerce, credit and cash in eighteenth-century advertisements', *Studies in Eighteenth-Century Culture*, 28 (1999), 103–130.
Conrad, Peter, *The Medicalisation of Society: On the Transformations of Human Conditions into Treatable Disorders* (Baltimore, MD: The Johns Hopkins University Press, 2007).
Cooter, Roger, *Surgery and Society in Peace and War: Orthopaedics and the Organisation of Modern Medicine, 1880–1948* (Basingstoke: Palgrave Macmillan, 1993).
Cooter, Roger and John Pickstone (eds), *Companion to Medicine in the Twentieth Century* (London: Routledge, 2003).
Corbin, Alain, *The Foul and The Fragrant: Odor and the French Social Imagination* (Cambridge, MA: Harvard University Press, 1996).
Corbin, Alain, *Time Design and Horror: Toward a History of the Senses*, trans. by Jean Birrell (Cambridge: Polity Press, 1995).
Corson, Richard, *Fashions in Eyeglasses*, 3rd edn (London: Peter Owen, 2011).
Crary, Jonathan, *Techniques of the Observer: On Vision and Modernity in the Nineteenth Century* (Cambridge, MA: MIT Press, 1990).
Crawforth, Michael A., 'Evidence from trade cards for the scientific instrument industry', *Annals of Science*, 42:5 (1985), 453–544.
Cryle, Peter and Elizabeth Stephens, *Normality: A Critical Genealogy* (Chicago, IL: University of Chicago Press, 2017).
Curtis, Ben and Steven Thompson, '"This is the country of premature old men": ageing and aged miners in the south Wales coalfield, c.1880–1947', *Cultural and Social History*, 12:4 (2015), 587–606.
Daston, Lorraine (ed.), *Things that Talk: Object Lessons from Art and Science* (New York: Zone Books, 2004).
Daston, L. and P. Gallison, *Objectivity* (New York: Zone Books, 2007).
Davidson, Derek C., 'Matthew William Dunscombe the first great collector of antique spectacles', *Ophthalmic Antiques Newsletter*, 4, 51.

Davidson, Derek C., *Spectacles, Lorgnettes and Monocles* (Princes Risborough: Shire, 2002).

Davidson, Luke, '"Identities ascertained": British ophthalmology in the first half of the nineteenth century', *Social History of Medicine*, 9:3 (1996), 313–333.

Davis, Lennard J., *Enforcing Normalcy: Disability, Deafness, and the Body* (London: Verso, 1995).

Digby, Anne, *The Evolution of British General Practice, 1850–1948* (Oxford: Oxford University Press, 1999).

Dixey & Son, C.W., *A Short History, 1777–1977* (Newport: Mullock & Sons, 1977).

Easley, Alexis, Andrew King and John Morton (eds), *Researching the Nineteenth-Century Periodical Press: Case Studies* (London: Routledge, 2018).

Elson, C.W., *Origin and Development of Spectacles* (Worthing: Worthing Archaeological Society, 1935).

Flint, Kate, *The Victorians and the Visual Imagination* (Cambridge: Cambridge University Press, 2000).

Forbes, Eric G., 'The professionalisation of dentistry in the United Kingdom', *Medical History*, 29 (1985), 169–181.

Foucault, Michel, *The Birth of the Clinic: An Archaeology of Medical Perception* (New York: Vintage Books, 1975).

Foucault, Michel, *The History of Sexuality*, vol. 1 (London: Penguin Books, 1990).

Garland-Thomson, Rosemarie, *Staring: How We Look* (Oxford: Oxford University Press, 2009).

Gates, Catherine, 'Matthew William Dunscombe: a Bristol optician', research paper (1997).

Gavrus, Delia, 'Men of dreams and men of action: neurologists, neurosurgeons, and the performance of professional identity, 1920–1950', *Bulletin of the History of Medicine*, 85 (2011), 57–92.

Goggins, Sophie, Tacye Phillipson and Samuel J.M.M. Alberti, 'Prosthetic limbs on display: from maker to user', *Science Museum Group Journal*, 8 (2017). DOI: 10.15180/170806 (accessed 28 March 2019).

Gooday, Graeme and Karen Sayer, *Managing the Experiences of Hearing Loss in Britain, 1830–1930* (Basingstoke: Palgrave Macmillan, 2017).

Gooding, Jo, 'Rather unspectacular: design choices in National Health Service glasses', *Science Museum Group Journal*, 7 (2017). DOI: 10.15180/170703.

Granshaw, Lindsay and Roy Porter (eds), *The Hospital in History* (London: Routledge, 1989).

'Guest editorial: hypermetropia or hyperopia?', *Ophthalmic & Physiological Optics*, 35 (2015), 2–7.

Guffey, Elizabeth and Bess Williamson (eds), *Making Disability Modern: Design Histories* (London: Bloomsbury, 2020).

Gurney, Peter, '"The age of veneer": Charles Dickens and the antinomies of Victorian consumer culture', *Dickens Quarterly*, 32:3 (2015), 229–246.

Gurney, Peter, *Wanting and Having: Popular Politics and Liberal Consumerism in England, 1830–1870* (Manchester: Manchester University Press, 2015).

Guyatt, M., 'Better legs: artificial limbs for British veterans of the First World War', *Journal of Design History*, 14:4 (2001), 307–325.

Hacking, Ian, 'Biopower and the avalanche of printed numbers', *Humanities in Society*, 5:3 and 4 (1982), 279–295.

Hamlin, Christopher, 'Review of John M. Eyler, *Sir Arthur Newsholme and State Medicine, 1885–1935*', *Bulletin of the History of Medicine*, 72:3 (1998), 564–566.

Hamling, Tara and Catherine Richardson (eds), *Everyday Objects: Medical and Early Modern Material Culture and its Meanings* (Farnham: Ashgate Publishing, 2010).

Handley, Neil, *Cult Eyewear: The World's Enduring Classics* (London: Merrell Publishers Ltd, 2011).

Harvey, Karen (ed.), *History and Material Culture* (Abingdon: Routledge, 2009).

Hatter, Janine and Nickianne Moody (eds), *Fashion and Material Culture in Victorian Fiction and Periodicals* (Brighton: Edward Everett Root, 2019).

Heath, Kay, *Aging by the Book: The Emergence of Midlife in Victorian Britain* (New York: State University of New York Press, 2009).

Hilaire-Pérez, Liliane and Christelle Rabier, 'Self-machinery? Steel trusses and the management of ruptures in eighteenth-century Europe', *Technology and Culture*, 54:3 (2013), 460–502.

Hiner, Susan, *Accessories to Modernity: Fashion and the Feminine in Nineteenth-century France* (Philadelphia: University of Pennsylvania Press, 2010).

Hirst, J.D., 'Vision testing in London: a rehearsal for the School Medical Service', *Journal of Education Administration & History*, 14:2 (1982), 23–29.

Holtman, H.W., 'A short history of spectacles', in W. Poulet, *Atlas on the History of Spectacles*, trans. Frederick C. Blodi (Godesberg: Wayenborgh, 1978).

Hutchison, Iain, Martin Atherton and Jaipreet Virdi (eds), *Disability and the Victorians: Attitudes, Interventions, Legacies* (Manchester: Manchester University Press, 2020).

Ilardi, Vincent, *Renaissance Vision from Spectacles to Telescopes* (Philadelphia, PA: American Philosophical Society, 2007).

Jay, Martin, 'The rise of hermeneutics and the crisis of ocularcentrism', *Poetics Today*, 9:2 (1988), 307–326.
Jones, Claire L., 'Instruments of medical information: the rise of the medical trade catalog in Britain 1750–1914', *Technology and Culture*, 54:3 (2013), 563–599.
Jones, Claire L., *The Medical Trade Catalogue in Britain, 1870–1914* (London: Pickering and Chatto, 2013).
Jones, Claire L. (ed), *Rethinking Modern Prostheses in Anglo-American Commodity Cultures 1820–1939* (Manchester: Manchester University Press, 2017).
Keeler, C. Richard, 'The ophthalmoscope in the lifetime of Hermann von Helmholtz', *Archives of Ophthalmology*, 120:2 (2002), 194–201.
Keeler, R., 'Antique ophthalmic instruments and books: the Royal College Museum; Part I Instruments', *British Journal of Ophthalmology*, 86:6 (2002), 602–603.
Khasnabis, Chapal, Zafar Mirza and Malcolm MacLachlan, 'Opening the gate to inclusion for people with disabilities', *The Lancet*, 386:10010 (2015), 2229–2230.
LaCom, Cindy, '"The time is sick and out of joint": physical disability in Victorian England', PMLA, 120:2 (2005), 547–552.
Larson, Frances, 'The things about Henry Wellcome', *Journal of Material Culture*, 15 (2010), 83–104.
Lawrence, Christopher, *Medicine in the Making of Modern Britain, 1790–1920* (London: Routledge, 1994).
Levine, Philippa, *The Amateur and the Professional: Antiquarians, Historians and Archaeologists in Victorian England, 1838–1886* (Cambridge: Cambridge University Press, 2002).
MacGregor, Arthur, 'Exhibiting evolutionism: Darwinism and pseudo-Darwinism in museum practice after 1859', *Journal of the History of Collections*, 21:1 (2009), 77–94.
Malleck, Daniel J., 'Professionalism and the boundaries of control: pharmacists, physicians and dangerous substances in Canada, 1840–1908', *Medical History*, 48:2 (2004), 175–198.
McGuire, Coreen, *Measuring Difference, Numbering Normal: Setting the Standards for Disability in the Interwar Period* (Manchester: Manchester University Press, 2020).
Meckel, Richard A., *Classroom and Clinic: Urban Schools and the Protection and Promotion of Child Health, 1870–1930* (New Brunswick, NJ: Rutgers University Press, 2013).
Mitchell, Margaret, 'Optics and the Science Museum', *The Optician* (21 September 1979), 22–24.
Mitchell, Margaret, *History of the British Optical Association, 1895–1978* (London: British Optical Association, 1982).

Morrison-Low, A.D., *Making Scientific Instruments in the Industrial Revolution* (Aldershot: Ashgate, 2007).
Mosley, Adam, 'Introduction [to special issue "Objects, texts and images in the history of science"]', *Studies in the History & Philosophy of Science*, 38 (2007), 289–301.
Mussell, James, 'Elemental forms: the newspaper as popular genre in the nineteenth century', *Media History*, 20:1 (2014), 4–20.
Nair, Aparna, '"They shall see his face": blindness in British India, 1850–1950', *Medical History*, 61:2 (2017), 181–199.
Nead, Lynda, 'The layering of pleasure: women, fashionable dress and visual culture in the mid-nineteenth century', *Nineteenth-Century Contexts*, 35:5 (2013), 489–509.
Nicolson, Malcolm, 'The emergence of orthodontics as a speciality in Britain: the role of the British Society for the Study of Orthodontics', *Medical History*, 51 (2007), 379–398.
Nutton, Vivian and Roy Porter (eds), *The History of Medical Education in Britain* (Amsterdam: Rodopi, 1995).
Nye, Robert A., 'The evolution of the concept of medicalization in the late twentieth century', *Journal of History of the Behavioural Sciences*, 39:2 (2003), 115–129.
Ogden, Daryl, *The Language of the Eyes: Science, Sexuality, and Female Vision in English Literature and Culture, 1690–1927* (New York: State University New York Press, 2005).
Oldenziel, Ruth and Mikael Hard, *Consumers, Tinkerers, Rebels: The People Who Shaped Europe* (Basingstoke: Palgrave Macmillan, 2013).
Ophthalmic Antiques Collectors Club Bulletin (East Chillington: The Club, 1982–1985).
O'Malley, Tom, 'The regulation of the press' in Martin Conboy and John Steel (eds), *Companion to British Media History* (London: Routledge, 2014), pp. 228–238.
Orr, Hugh, *Illustrated History of Early Antique Spectacles* (London: author, 1985).
Ott, Katherine, David Serlin and Stephen Mihm (eds), *Artificial Parts and Practical Lives: Modern Histories of Prosthetics* (New York: NYU Press, 2002).
Otter, Chris, *The Victorian Eye: A Political History of Light and Vision in Britain, 1800–1910* (Chicago, IL: University of Chicago Press, 2008).
Oudshoorn, Nelly and Trevor Pinch (eds), *How Users Matter: The Co-Construction of Users and Technology* (Cambridge, MA: MIT Press, 2005).
Park, Hyung Wook, *Old Age, New Science: Gerontologists and Their Biosocial Visions, 1900–1960* (Pittsburgh, PA: University of Pittsburgh Press, 2016).

Parr, Joy, *Sensing Changes: Technologies, Environments and the Everyday, 1953–2003* (Vancouver: UBC Press, 2010).
Perkin, H., *The Rise of Professional Society: England since 1880* (London: Routledge, 2002).
Phillips, Gordon, *The Blind in British Society: Charity, State and Community c.1780–1930* (Aldershot: Ashgate, 2004).
Porter, Roy, *Health for Sale: Quackery in England, 1660–1850* (Manchester: Manchester University Press, 1989).
Qureshi, Sadiah, *People on Parade: Exhibitions and Anthropology in C19th Britain* (Chicago, IL: University of Chicago Press, 2011).
Reinarz, Jonathan, *Past Scents: Historical Perspectives on Smell* (Urbana: University of Illinois Press, 2014).
Reiser, Stanley Joel, *Medicine and the Reign of Technology* (Cambridge: Cambridge University Press, 1978).
Reiss, Matthias, *Blind Workers against Charity: the National League of the Blind of Great Britain and Ireland, 1893–1970* (Basingstoke: Palgrave Macmillan, 2015).
Richards, N. David, 'Dentistry in England in the 1840s: the first indications of a movement towards professionalisation', *Medical History*, 12 (1968), 137–152.
Roach, J.P.C., *Public Examinations in England, 1850–1900* (Cambridge: Cambridge University Press, 1971).
Robinson, Emily, 'Touching the void: affective history and the impossible', *Rethinking History: The Journal of Theory and Practice*, 14:4 (2010), 503–520.
Rose, Sarah F., *No Right to be Idle: the Invention of Disability, 1840s–1930s* (Chapel Hill: The University of North Carolina Press, 2017).
Rosen, George, *The Specialisation of Medicine with Particular Reference to Ophthalmology* (New York: Froben Press, 1944).
Rosenthal, William, *Spectacles and Other Vision Aids: A History and Guide to Collecting* (San Francisco, CA: Norman, 1996).
Segrave, Kerry, *Vision Aids in America: A Social History of Eyewear and Sight Correction since 1900* (London: McFarland & Company Inc., 2011).
Shortt, S.E.D., 'Physicians, science and status: issues in the professionalisation of Anglo-American medicine in the nineteenth century', *Medical History*, 27 (1983), 51–68.
Shuttleworth, Sally and Geoffrey Cantor (eds), *Science Serialized: Representations of the Sciences in Nineteenth-Century Periodicals* (London: The MIT Press, 2004).
Skinner, Ghislaine M., 'Sir Henry Wellcome's Museum for the Science of History', *Medical History*, 30 (1986), 383–418.
Smith, Mark M., *A Sensory History Manifesto* (Philadelphia: Pennsylvania State University Press, 2021).

Stark, James, *The Cult of Youth: Anti-Ageing in Modern Britain* (Cambridge: Cambridge University Press, 2020).

Stobart, Jon, 'Selling (through) politeness: advertising provincial shops in eighteenth-century England', *Cultural and Social History*, 5:3 (2008), 309–328.

Stobart, Jon, Andrew Hann and Victoria Morgan, *Spaces of Consumption: Leisure and Shopping in the English Town, c.1680–1830* (London: Routledge, 2007).

Sweet, Ryan, *Prosthetic Body Parts in Nineteenth Century Literature and Culture* (London: Palgrave Macmillan, 2022). Open access: DOI: 10.1007-978-3-030-78589-5.

Taylor, Geoffrey Stuart and Malcolm Nicolson, 'The emergence of orthodontics as a speciality in Britain: the role of the British Society for the Study of Orthodontics', *Medical History*, 51(2007), 379–398.

Thane, Pat, *Old Age in English History: Past Experiences, Present Issues* (Oxford: Oxford University Press, 2002).

The Newsletter: Ophthalmic Antiques International Collectors Club (East Chillington: The Club, 1985 to present).

Timmermann, Carsten and Julie Anderson (eds), *Devices and Designs: Medical Technologies in Historical Perspectives* (Basingstoke: Palgrave Macmillan, 2006).

Turner, David M. and Alun Withey, 'Technologies of the body: polite consumption and the correction of deformity in eighteenth-century England', *History: The Journal of the Historical Association*, 99:338 (2014), 775–796.

Turner, David M., 'Disability history and the history of emotions: reflections on eighteenth-century Britain', *Asclepio*, 68.2 (2016). DOI: 10.3989/asclepio.2016.18.

Turner, David M. and Daniel Blackie, *Disability in the Industrial Revolution: Physical Impairment in British Coalmining, 1780–1880* (Manchester: Manchester University Press, 2018).

Ueyama, Takahiro, *Health in the Marketplace: Professionalism, Therapeutic Desires, and Medical Commodification in Late-Victorian London* (California: SPSS, 2010).

Ugolini, Laura and John Benson (eds), *Retailing Beyond the Shop, c.1400–1900* (Bradford: Emerald Group Publishing Ltd, 2010).

Underwood, E.A. (ed.), *Science, Medicine and History: Essays on the Evolution of Scientific Thought and Medical Practice Written in Honour of Charles Singer* (Oxford: Oxford University Press, 1954).

Virdi-Dhesi, Jaipreet, 'Curtis' cephaloscope: deafness and the making of surgical authority in London, 1816–1845', *Bulletin of the History of Medicine*, 87:3 (2013), 347–377.

Virdi-Dhesi, Jaipreet, 'From the hands of quacks: aural surgery, deafness, and the making of a surgical specialty in 19th century London' (2014), unpublished PhD thesis, University of Toronto.

Virdi, Jaipreet and Coreen McGuire, 'Phyllis M. Tookey Kerridge and the science of audiometric standardisation in Britain', *British Journal for the History of Science*, 51:1 (2018), 123–146.

Vogel, Wolfgang H. and Andreas Berke, *Brief History of Vision and Ocular Medicine* (Amsterdam: Wayenborgh Publishers, 2009).

Waddington, Keir, 'Mayhem and medical students: image, conduct, and control in the Victorian and Edwardian London teaching hospital', *Social History of Medicine*, 15:1 (2002), 45–64.

Wallis, Patrick, 'Consumption, retailing and medicine in early modern London', *Economic History Review*, 61:1 (2008), 26–53.

Wallis, Patrick, 'Introduction: the growth of the early modern medical economy', *Journal of Social History*, 49:3 (2016), 477–483.

Waters, Catherine, '"Fashion in undress": clothing and commodity culture in *Household Words*', *Journal of Victorian Culture*, 12:1 (2007), 26–41.

Wear, Andrew (ed.), *Medicine in Society* (Cambridge: Cambridge University Press, 1992).

Weaver, Caroline, 'Eyesight and governance in Britain: bureaucracy and the senses in the 1920s', *Social History of Medicine*, 27:2 (2014), 241–259.

Weisz, George, 'The emergence of medical specialisation in the nineteenth century', *Bulletin for the History of Medicine*, 77 (2003), 536–575.

Weisz, George, *Divide and Conquer: A Comparative History of Medical Specialisation* (New York: Oxford University Press, 2006).

Williams, Kevin, *Read All About It! A History of the British Newspaper* (London: Routledge, 2010).

Williamson, Bess, 'Electric moms and quad drivers: people with disabilities buying, making, and using technology in postwar America', *American Studies*, 52:1 (2012), 5–30.

Willis, Martin, *Vision, Science and Literature, 1870–1920: Ocular Horizons* (London: Pickering & Chatto, 2011).

Withey, Alun, *Technology, Self-Fashioning and Politeness in Eighteenth-century Britain: Refined Bodies* (Basingstoke: Palgrave Macmillan, 2016).

Woolgar, Christopher Michael, *The Senses in Late Medieval England* (New Haven, CT; London: Yale University Press, 2006).

World Health Organization, *World Report on Vision* (Geneva: World Health Organization, 2019).

Zwierlein, Anne-Julia, Katharina Boehm and Anna Farkas (eds), *Interdisciplinary Perspectives on Aging in Nineteenth-Century Culture* (New York: Routledge, 2014).

Websites

Definition of disability under the Equality Act 2010 – GOV.UK (archive.org) (accessed: 23 April 2022).

Priority Assistive Products List (APL) (who.int) (accessed 15 February 2022).

Thackray Medical Museum's Collections Online | Home (archive.org) (accessed: 3 February 2022).

Index

accessories 4, 30, 42, 138, 208, 209, 223, 226, 228, 229, 231, 232–233, 247
advertisement 9, 26, 28, 29, 38n.83, 41–42, 43, 45–48, 51–53, 57, 60–62, 64–65, 70, 104, 124, 129, 130, 134, 136–138, 143, 146–149, 164, 166, 170, 171–172, 175–176, 177, 179, 180, 187, 190, 193, 194, 211, 226, 228
  false 57, 61–62, 143
  puffing 41, 62
age 2, 30, 41, 58, 69, 86, 110, 153, 165, 209, 210–211, 220, 222–225, 227, 228, 235–237, 250
ageing 85–86, 88, 209, 211, 222–225
  stigma and 30, 221–224
Aitchison, James 136, 143, 145, 191–192
alternatives to spectacles 28, 40, 66–68, 221
America 10, 45, 56, 58, 63, 76n.70, 76n.78, 77n.93, 77n.96, 83, 90–91, 96, 98, 99, 101, 105, 107, 108, 127, 141, 233, 231, 240n.43, 240n.54, 253
  Anglo-American 218

Anderson, Julie 165, 193, 211
army 107–108, 120n.117, 121n.118
artificial limbs 14, 26, 144, 165, 166, 193, 198n.21, 211, 213–214, 222
  see also prosthetics
assistive device/technology 42, 55, 57, 140, 165, 207, 246
  assimilation of 21–22, 29–30, 144–5, 152, 167, 197, 210–211, 222, 234, 236, 252–253
  attitudes to 30, 125–126
  definitions of 14–16
  disguise and 223
  invisible 20, 226–228, 235–236, 252
  market 14, 20, 164–165, 236
  shaping the meaning of disability 16, 18–19, 21–22, 31, 207–210, 213–214, 217–220, 226, 236, 254
asthenopia 67, 79n.123, 92, 213
astigmatism 16, 67, 85, 93, 163, 182, 215
aurist 90, 104
average 9, 82–83, 86, 88–89, 152, 220, 251

Index

bazaar 40, 44, 47, 68, 247
beauty 20, 209, 222, 228
Beer, Georg 58
blindness 7–8, 177, 221, 224, 250, 253
   changing definitions of 16–17, 19, 28, 30, 207, 210–211, 212–215, 219–220, 236, 255
Board of Trade 109, 111, 113
Booth, Charles 143–144, 159n.60
British Medical Association 94, 187, 217
British Medical Journal see publications
British Optical Association (BOA) 59, 186, 189–190, 195–196, 235
   see also College of Optometrists
Brownlee, Peter John 10, 45, 63, 76n.70, 76n.78, 77n.93, 77n.96, 87, 119n.95, 239n.40, 240n.43, 240n.54
'economy of the eyes' 10, 103
built environment 2, 5, 10, 12
   see also urban and urbanisation

Canguilhem, Georges 83
Carter, Robert Brudenell 97–98, 99–100, 105, 132, 138–139, 142, 151, 173, 216–217, 219
Chadburn and Son 141–142, 143, 145, 146, 152, 235
children 2, 96–102, 119n.90, 142, 149, 214, 217, 221, 223
class *see* social, class
College of Optometrists 38n.85, 186
   see also British Optical Association
Conrad, Peter 13, 14, 102
Cooper, William White 104, 212–213, 216
corrective body devices 40, 43, 57, 70

Cryle, Peter and Elizabeth Stephens 18, 82–83, 89
cure *see* medical, cure
Curry & Paxton 143, 145, 169
   see also Pickard and Curry 169, 171
Curtis, John Harrison 55, 116n.38, 119n.98, 104

Davis, Lennard J. 17, 82, 88–89
deafness 7, 8, 17, 89, 95, 214
   see also hearing, loss
degeneration 96, 225–226
design
   model of disability 21–22, 25, 209
   shaping contemporary meaning and attitudes to blindness or partial sight 31, 126, 207–211, 228, 235–237, 249–250, 252, 254–255
   visual aids and 10, 23–24, 29, 50, 123–141, 145, 148, 152–153, 193, 197, 210–211, 236–237, 246–247, 250–252
diagnostic
   equipment 3, 14, 59, 84–85, 92–93, 170, 172, 185
   instrument 64, 90
   technology 3–4, 17, 28, 66, 84, 89–90, 169, 214, 249, 255
   tools 80, 82, 249
dioptre 93, 172
dioptric 189, 195–196
disability
   history of 2, 4, 7, 8, 31, 89, 248
   shifting definitions of 15–22, 82, 88–90, 92, 112, 125, 208–220, 249–250, 252, 254–255
   industrialisation and 107, 109, 226
   material culture of 22–23
   passing and 226, 236

Index

Dixey, C.W. 68–69, 144–145, 167, 168, 185–186, 228
Donders, Franciscus Cornelis 17, 84–88, 91–92, 170, 213, 224, 230
Dunscombe, Matthew 24–25, 26, 123–124, 130, 135–136, 138, 140, 141, 146

education 2, 5, 8, 9, 15, 95–98, 100–101, 151, 173, 184, 220–221, 225, 232
see also schools
Education Act (1870) 96, 97
Europe 1, 58, 83, 96, 98, 108, 109, 110, 141, 253
Austria 58, 223
European 9, 87, 111, 125
France 99
Germany 80, 98–99, 105, 141, 176, 180
Netherlands 17, 84
Switzerland 91, 45

Fallon, Cara Kiernan 21, 208–209
fashion 5, 6, 20, 208, 228, 230–235
fashioning
the body 5, 125, 208, 228
definition of 210
the face 30, 207, 209, 236
vision 212, 214, 216, 220
Foucault, Michel 12

gender 6, 30, 209–210, 211, 222, 224–226, 228, 230–234, 236–237
Gooday, Graeme and Karen Sayer 7
Guffey, Elizabeth and Bess Williamson 21, 22, 25, 237n.9

Harris & Son, Thomas 9, 51, 53, 55, 56, 62, 158n.52, 222, 242n.90

hearing
aid 14, 16, 26, 165, 235–236, 252
loss 82, 89, 212
categories of 89, 212, 213–214
Helmholtz, Hermann von 80, 84–85
hospital 4, 12, 57, 58, 65, 70, 94–95, 104, 106, 151, 153–154, 163–164, 167, 169, 171, 172, 180, 181, 189, 192, 193, 196, 212, 215, 216, 217
Horner, Friedrich 91, 145–146
Hudson, John Thomas 39–42, 53, 56–57, 63, 132
hypermetropia 16, 38n.83, 85–86, 92, 93, 213
see also long sight

itinerant pedlar 44, 50, 51, 52, 247
see also spectacle hawker and street seller
industrialisation 44, 103–104, 107, 110, 226
steam 125
factories 141–143
intersectionality 247

jeweller 40, 42, 43, 45, 46–47, 68, 106, 146, 148, 150, 166, 175, 176, 178, 179–180, 187–189, 207
Johnson, Lindsay 189, 191–192
Jones, Claire L. 19–20, 38n.81, 78n.61, 125, 164, 170, 174, 208, 210

Lancet, The see publications
lenses 6, 9, 11, 16, 19, 67, 125, 138
bifocal 138, 159n.74
coloured 8, 128, 236

lenses (*cont.*)
  concave 41, 67, 126–127, 169, 181, 213, 224–225
  convex 41, 126–127, 150, 169, 181, 224
  cylindrical 138, 142
  glass 3, 28, 55, 142–143, 169, 179, 212
  pebble 127–128, 132, 142, 145, 180
  protective 9, 128
  tinted 150
  trial 42, 57, 62–64, 172–6, 92–93, 163, 167–168, 170, 173
  utility of 19, 126–127, 195, 210, 212–220, 236, 249
  see also manufacture, lenses
long sight 16, 85, 105, 127
  see also hypermetropia
Long, Charles 55–56, 68, 127

Mackenzie, William 66–67, 146, 213, 215
manufacture 6, 10, 29, 39, 43, 48, 53, 57, 153
  spectacle frame 11, 15, 46, 47, 124–127, 139–145, 226–227
  lens 39, 53, 126, 138, 141–145, 185
  testing equipment 167–168, 170–172
mass manufacture 21, 29, 125–126, 140–141, 142, 151–153, 247
material culture 22–27, 28, 29, 30, 42, 124
  collectors 5–7, 22–25, 37n.74, 124, 180
  objects 2, 6–7, 14, 19, 40, 43, 70, 72n.27, 78n.102, 124, 155n.18, 198n.14, 199n.26, 199n.29, 242n.89–90, 254
  touch 23, 124, 254
  agency and 23, 231–232

  see also Science Museum's Collections and disability, material culture of
Mayhew, Henry 50–51, 61
McGuire, Coreen 17, 36n.55, 82, 89, 212, 250
mechanical objectivity 82, 89–90
  see also objective testing
medical
  capitalism 5, 191
  'cure' 12–13, 18–19, 67, 214, 220, 249
  instruments *see* diagnostic, equipment/ instrument/ technology/ tools
  general practitioner 58, 219
  marketplace 13–14, 42, 164–165, 177, 193
  model of disability 19, 21, 89, 174
  orthodoxy 167
  specialisation 4, 11, 40, 58, 81, 183
  specialty 13, 94, 96–97
  specialism 90, 112–113, 251
  specialists 170, 174
  technology *see* diagnostic, equipment/ instrument/ technology/ tools
  trade catalogues 26, 28, 75n.61, 83, 92–93, 164, 169–171
medicalisation 4, 5, 11–15, 19–20, 95, 102, 214, 254
military *see* army
Mitchell, Margaret 59
Moorfields Eye Hospital 58, 151, 217–219, 221
  see also hospital
myopia 13, 16, 38n.83, 55, 67, 85–86, 96, 101, 106, 151, 164, 176, 216, 218, 223, 225
  see also short sight

Index

National Health Service (NHS) 251–252
normal vision 12, 17–18, 21–22, 28, 29–31, 82–84, 86–89, 102–103, 106, 109, 111, 112–113, 125–126, 152–153, 175, 250, 251–253
  see also vision, norms of
objective testing 41–42, 59, 64–65, 81–82, 89–91, 92, 94, 102, 108, 112, 249, 250, 255
  see also mechanical objectivity
oculist 13, 38n.83, 41, 58, 104, 105, 109, 167, 170, 171–172, 174–175, 178, 181–182, 184, 185, 186, 187, 191, 216
'old sight' 41, 55, 67, 85–86, 126
  see also presbyopia
orthopaedics 11, 81, 190
ophthalmologist 3, 4, 11–15, 17, 28–30, 40, 42, 58, 64, 65–68, 81, 83, 84–85, 87, 90–102, 104–109, 111–113, 127, 132, 138, 145, 151, 164–177, 183–186, 189–196, 207, 212, 216–219, 223, 224, 230, 250–251
ophthalmology 1, 11–15, 28, 40, 42, 57–58, 65, 69–70, 81, 94–95, 98, 103, 104, 184–185
  history of 11, 180
  specialist eye institution *see* hospital
ophthalmoscope 3, 12, 14, 17, 28, 80–81, 82, 84–85, 87, 90–91, 92–93, 102, 163, 164, 166, 169, 170, 213, 222, 254
opticians
  certification of 15, 182, 186–191, 192, 195–196, 250–251
  definition of 39 159n.60, 166, 176
  itinerant 45–46, 52, 180
  pamphlet 39, 51, 138, 164, 175, 176, 177–178
  professional role 3, 4, 14–15, 28, 29–30, 40, 43, 56–57, 58–60, 69–70, 164, 166, 176, 179, 182–197, 217–219, 248
  retailing practices of 43–70, 170–173, 174, 179–181
  text 124, 130, 142, 178
  treatise 42, 51, 55, 56, 62
Optician, The see publications
optometer 64, 66, 68, 92, 108, 169, 194
optometrist 1, 166, 186, 196, 250, 251
optometry 11–13, 59, 180
optotypes 80, 87, 163
  see also test-types
Ott, Katherine 15–16, 22, 25
Otter, Chris 7, 20, 100

patent 14, 26, 42, 124, 135–136, 137, 138, 139, 140–141, 142, 165, 181, 222, 226
pince-nez 137, 138, 152, 157n.35, 231
presbyopia 67, 85–86, 253
  see also 'old sight'
professional identity 11, 30, 68, 70, 90, 94–95, 165, 167, 178, 179, 182, 183–184, 185–186, 191, 196, 224, 251
  boundaries 3, 13, 15, 26, 29, 165, 166, 167–168, 182, 184–186, 190–191, 199n.34, 250–251
professionalisation 4, 81, 98, 173–174, 183, 194
prosthetics 7, 14–16, 20, 23, 38n.81, 148, 193, 226, 222, 231, 253
  invisible and 226–228

prosthetics (*cont.*)
  disguise and 222, 231
  appearance and 148, 161n.83,
    234
  commodification of 14, 125,
    148–150, 165, 174
  market 19, 164–165
  shaping meaning of disability 17,
    19–20, 125–126, 208, 210,
    213
publications
  British Medical Journal 26, 93,
    95, 96–97, 100, 107–108,
    109–112, 168–169, 177,
    189–192, 196
  Lancet, The 183, 185–186, 189
  Optician, The 26, 29, 127,
    134–135, 136, 138, 139, 141,
    143–145, 174–176, 179, 180,
    181–190, 193, 235, 236

quackery 12, 13, 27, 58–59, 61–62,
    94, 188
  charlatan 191
  quack 41–42, 52, 175, 179,
    193–194
Quetelet, Adolphe 88

railway 108–113, 181, 196
Reiser, Stanley 88, 119n.94
retailers 4, 13–15, 28, 40–48, 51,
    53, 57, 59, 62, 63–64, 65, 70,
    147–148, 151, 166, 183, 194,
    196, 226, 228, 248, 249
  high street 4, 15, 29–30, 44, 164,
    167, 169–171, 175, 179, 181,
    183, 192, 196–197, 207, 250
  reputation and 43, 52, 53,
    203n.82, 193

scientific instrument
  trade 28, 40, 44, 50, 54–55,
    143

maker 6, 40, 43–44, 45, 48, 49,
    64, 166, 179
  schools 28, 83, 95–102, 105, 106,
    100, 110, 111, 113, 149, 152,
    174, 176, 214, 224, 229–230
  see also education and Education
    Act (1870)
Science Museum's collections
    23–28, 43, 46, 47, 124, 129,
    48, 63, 130–131, 133, 140,
    141, 142, 149, 164, 167,
    171, 172, 181, 226, 228–229,
    257
sensory
  capacity 2, 14, 16, 18, 165, 212,
    213, 215–216, 249–250
  hierarchy 7–8, 248
  history 8, 31, 248–249
  short sight 2, 3, 13, 16, 41, 55, 66,
    69, 85–86, 98, 101, 105, 106,
    199n.88, 127, 163, 176,
    209–210, 211, 212–213, 215,
    225, 227–228, 232, 233–234
  see also myopia
Smee, Afred 65–66, 68, 104
  Smee's optometer 68
Smith, Martin M. 8, 248
Snell, Simeon 97, 98, 217
Snellen, Herman 87–88, 91
social
  class 1, 26, 30, 52, 103, 148,
    153, 161n.83, 163, 164,
    210–211, 222, 227–228,
    229–231, 233–235, 236–237
  model of disability 89
spectacle hawker 50, 52, 56, 59–60,
    180, 181, 194, 196
spectacle maker 44, 52, 55, 143,
    145, 159n.60
Spectacle Makers' Company
  (SMC) 4, 186–192, 195
see also itinerant pedlar and
    street seller

Index

spectacle wear
  increase of 2–3, 8–9, 10–11, 102, 118n.77, 119n.90, 124–125, 151–152, 220–221, 222–223
    see also vision, deterioration of
  standardisation 4, 18, 82, 87–89, 91–92, 93, 101, 107, 110, 113, 125, 130–131, 140, 144–145, 166, 209, 215, 216, 250
  statistics 3, 17–18, 81–82, 88–89, 97–98, 101, 105, 108–109
steel 14, 23, 42, 45, 47, 125, 128, 130–131, 133–135, 137, 140–141, 146, 165, 171, 235, 236–237
stigma 30, 209, 210–211, 220, 224, 225–226, 233–234, 235–237, 252
street seller 50–52, 60–61
  see also itinerant pedlar and street seller
Sweet, Ryan 161n.83, 222, 225, 234

test-types 91–92, 93, 116n.43, 173
  see also optotypes
trade cards 26, 28, 48–49, 53–54, 65
trade catalogues 48, 63, 145–146
  see also medical, trade catalogues
trade directory 26, 44, 55, 171, 180, 191
Turner, David M. 8, 70

Ueyama, Takahiro 14, 164–165, 175, 177, 184, 190, 193
urban 8–10, 181, 209, 219

urbanisation 31, 247
  see also built environment

Virdi-Dhesi, Jaipreet 90, 119n.98
Virdi, Jaipreet 36n.55, 95, 214, 245n.134
vision
  deterioration 2–3, 8, 45, 95, 97, 101–102, 118n.77, 119n.90, 176, 177, 220–221, 225–226
  measurement of 4, 11, 13, 16, 27–28, 31, 65–66, 80–83, 87–91, 96, 100, 102, 103, 106–108, 112–113, 166, 215, 248
  norms of 1, 4–5, 12 18, 20, 81–84, 90, 112–113, 251
  testing 2, 3–4, 11, 17–18, 26, 28–29, 39–40, 41, 43, 59, 63–65, 80–84, 87–88, 91–95, 99–102, 107–113, 125, 129, 149, 154, 163–178, 181–186, 188–191, 192–196, 247, 212, 215, 216–217, 249, 250–251, 253

walking stick 15, 21, 22, 48, 209, 226
watchmaker 40, 43, 47, 105–106, 207
Wellcome, Henry 24–25, 26, 76n.69, 124, 146
West, Francis 55, 56, 61–62
Withey, Alun 38n.81, 70, 125–126, 209, 228, 236, 247
work 10, 19, 30, 61, 96, 212, 216, 218–220, 221, 225, 235
  Employer's Liability Act (1880) 109

work (cont.)
  environment 18, 83–84,
    102–113
  Workmen's Compensation Act
    (1897) 109, 110
  workers' bodies 5, 103, 106–107,
    109

workplace 5, 28, 83, 95, 97,
  102–104, 105–107, 109,
  113, 152, 174, 221, 250,
  252
  see also work, environment

Young, Thomas 64

EU authorised representative for GPSR:
Easy Access System Europe, Mustamäe tee 50,
10621 Tallinn, Estonia
gpsr.requests@easproject.com

www.ingramcontent.com/pod-product-compliance
Lightning Source LLC
Chambersburg PA
CBHW051603230426
43668CB00013B/1964